Contributions to Management Science

More information about this series at http://www.springer.com/series/1505

Azar Karimov

Identifying Stock Market Bubbles

Modeling Illiquidity Premium and Bid-Ask Prices of Financial Securities

Springer

Azar Karimov
Institute of Applied Mathematics
Middle East Technical University
Ankara, Turkey

ISSN 1431-1941 ISSN 2197-716X (electronic)
Contributions to Management Science
ISBN 978-3-319-65008-1 ISBN 978-3-319-65009-8 (eBook)
https://doi.org/10.1007/978-3-319-65009-8

Library of Congress Control Number: 2017949446

Printed on acid-free paper

This Springer imprint is published by Springer Nature
The registered company is Springer International Publishing AG
The registered company address is: Gewerbestrasse 11, 6330 Cham, Switzerland

To my family

Foreword

The purpose of this book, *Identifying Stock Market Bubbles: Modeling Illiquidity Premium and Bid-Ask Prices of the Financial Securities*, is to deeply inquire, holistically reflect on, and practically expose the current and emerging concept of modern time modeling to the areas of financial market dynamics and derivative pricing with real-time data and signals. During the previous decades, the analytical tools and the methodological toolbox of applied and financial mathematics, and of statistics, have gained the attention of numerous researchers and practitioners from all over the world, providing a strong impact also in economics and finance. Here, the notions of information on asset valuation and disparities among market participants are turning out to be key issues from an integrated perspective, and they are closely connected with further areas such as behavioral analysis, financial signal processing, agent-based modeling, and early detection of financial bubbles and liquidity squeezes.

This book seeks to revitalize, amid ongoing structural problems in financial markets, and reassess the role of liquidity through a new fundamental approach to pricing of a range of financial contracts, option contacts as a special case, which focuses on a novice, yet intuitive and theoretically robust, framework of illiquidity premium, with the help of the *Conic Finance Theory*. The book could become a helpful compendium for risk managers, decision-makers, researchers, as well as graduate students and practitioners in quantitative finance who aim to go beyond conventional approaches to financial modeling.

The author of this book is both an academic and practitioner in the area of applied financial mathematics, with considerable international research experience. He uses the state-of-the-art model-based strong methods of mathematics as well as the less model-based, more data-driven algorithms which are less rigorous mathematically and released from firm calculus in order to integrate data-led approaches with a view to efficiently coping with hard problems. Today, labeled by names like Conic Finance and Two Price Economic Valuation, and by Operational Research and Analytics, model-free and model-based streamlines of traditions and approaches meet and exchange in various centers of research, at important congresses, and in leading projects and agendas in all over the world. The herewith joint intellectual enterprise aims to benefit from synergy effects, to commonly advance scientific progress and to provide a united and committed service to the solution of urgent real-life challenges.

To the author of this valuable book, Dr. Azar Karimov (also known as Azer Kerimov), I extend my heartily appreciation and gratitude for having shared his devotion, knowledge, and vision with the academic community and mankind. I am very thankful to the publishing house *Springer*, and the editorial team around *Dr. Christian Rauscher* thereat, for having ensured and made become reality a premium work of a high-standard academic and applied importance, and a future promise of a remarkable impact for the world of tomorrow.

Now, I wish all of you a lot of joy in reading this interesting work, and I hope that a great benefit is gained from it both personally and societally.

Middle East Technical University
Ankara, Turkey
June 2017

Gerhard-Wilhelm Weber

Acknowledgements

I would like to thank my supervisor, Prof. Dr. Gerhard-Wilhelm Weber. It was a great honor to be his PhD student. Prof. Dr. Weber has given a gentle and benevolent guidance during the whole period of PhD Studies. I am very grateful for all of his contributions of time and ideas to make my PhD experience productive and stimulating. Without his support this PhD thesis would not have been written.

I would like to offer my sincere thanks to my co-supervisor Prof. Dr. Suat Teker for his invaluable suggestions. I benefited greatly from many fruitful discussions with him.

I owe my deepest gratitude to my father, Prof. Dr. Nazim Kerimov. Only with his warm encouragement, valuable guidance, and technical support this thesis have been finished. His help is comprehensive and his unstinting counsels and time made easier to write this thesis.

My greatest thankfulness goes to Dr. Efsun Kürüm who always gave a great support before and during the writing of this thesis.

I also would like to thank my friends Dr. Ayşe Özmen and Dr. Erdem Kılıç who always willingly and cordially helped me.

I would like to thank the administrative staff and all other members of the Institute of Applied Mathematics at Middle East Technical University for their patience and aids during my studies.

I am deeply grateful to my family for their endless love, patience, and encouragement throughout my life.

Contents

About the Author

Dr. Azar Karimov, CFA, FRM is the graduate in Financial Mathematics from the Institute of Applied Mathematics at Middle East Technical University. He has worked as a risk manager in intergovernmental diplomatic organization, Turkish private banking institutions and accumulated an extensive industry experience in liquidity management, financial risk management, stress testing, and asset-liability management. He has also delivered on-the-job trainings on advanced financial risk modeling at Turkish regulatory authorities.

List of Abbreviations

CDF	Cumulative Distribution Function
CDO	Collateralized Debt Obligation
CRLB	Cramer-Rao Lower Bound
DED	Double-Exponential Distribution
ECF	Empirical Characteristic Function
GMM	Generalized Method of Moments
i.i.d.	Independent and Identically Distributed
LPPL	Log-Periodic Power Law
MLE	Maximum-Likelihood Estimation
$\mathbb{N}$	Set of Natural Numbers
PC	Personal Computer
PDF	Probability Distribution Function
$\mathbb{R}$	Set of Real Numbers

List of Figures

List of Tables

Chapter 1
Introduction

Liquidity is the crucial prerequisite for the normal working of financial markets. Only the properly liquid money markets can work successfully. The liquidity of business sectors, more accurately the absence of it, influences the entire financial system, and the entire economy, repressing their typical way of working. The financial bubble of 2008 has indicated remarkable significance of the liquidity of the financial markets, and in the meantime it pushed this inquiry to the spotlight. The modification and supplementation of the standard equilibrium and no-arbitrage models which accept the presence of infinity liquidity has turned into a need, as there is a clear necessity to develop new pricing and detection models and risk management systems.

A complete description and basic interpretation of liquidity does not exist. Straightforward definitions in one sentence would be like, e.g., "Liquidity in a financial market is the capacity to assimilate easily the flow of buy and sell orders—. . ." [1]. Unfortunately, this definition cannot catch the reality of "liquidity", since liquidity is not a one-dimensional measurement but it rather incorporates several measurements.

In illiquid markets, sales charges are altogether higher than in liquid markets, i.e. deals may be executed with particularly higher expense and time. In this way it is not unexpected that investors' essential prerequisite is that every stock's liquidity ought to be proportionate and the trade costs have to be quantifiable. Measuring liquidity is a complicated issue by itself, it is hard to express the majority of its angles with one single pointer; likewise it is difficult to gauge how much cost illiquidity produces during the transaction, since liquidity can be explained along various measurements and accordingly at any given time some of its distinctive characteristics can be in the spotlight.

A. Karimov, *Identifying Stock Market Bubbles*, Contributions to Management Science, https://doi.org/10.1007/978-3-319-65009-8_1

Different aspects of liquidity may be organized as in below in order to display the given levels of liquidity:

1. *The capacity to make an exchange at all*: This first level of liquidity is self-evident: if there is no liquidity entirely in the market, no exchange will occur. In a liquid market, there exist no less than one bid and one ask quote that make an exchange conceivable.
2. *The capacity to buy or to sell a specific amount of an asset with impact on the given price*: In a liquid market, it is conceivable to exchange a specific measure of shares with little effect on the cited cost.
3. *The capacity to buy or to sell a specific quantity of an asset without impact on the cited price*: The more liquid a market, the less is the effect on the cited prices. Along these lines, as the liquidity rises, in the long run a point will be achieved where there is no more value effect for a specific quantity of shares.
4. *The capacity to buy and to sell an asset at about the same cost at the same time.*
5. *The capacity to execute a trade from items 2–4 quickly.*

During this research we put strong emphasis on an illiquidity indicator which evaluates the differences between the bid and ask prices of financial derivatives in a real time frame. Another principle objective of the thesis is to help illiquidity as an idea to be integrated into the routine of Risk Management Departments on a daily basis, i.e., to give detailed resolutions which can be effectively incorporated into their regular work, and also to correctly develop from hypothetical perspective. Moreover, by evaluating the illiquidity premium we try to detect the financial bubbles before it collapses.

Stock market bubbles take place when the prices of stocks are exaggeratingly higher than their *fundamental value* (*intrinsic value*). According to the fundamental value of a security, an investor can decide whether the security is *overvalued* or *undervalued*. If the intrinsic value of a security is higher than its current market price, it is classified as *overvalued*, if otherwise occurs (i.e. intrinsic value lower than market price), it is called as *undervalued*. Hence, an investor may build their investment decision based on this criteria.

The thesis is divided into eight chapters and is broadly focused on one major related topic, i.e., bid-ask spread or illiquidity premium. This chapter serves as an introduction to the thesis and provides various motivations for the research undertaken.

Chapter 2 provides a critical review of the current literature with existing bid-ask spread measurement approaches. For the purpose of this thesis, we classify the literature into three broad categories: (1) the first one includes the literature about the methods and techniques used in inventory models related to the bid-ask spread; (2) the second category consists of the literature about models and useful tools to extract the effective bid-ask spreads from information models; and (3) the third category focuses on the model, of Conic Finance Theory which studies liquidity or bid-ask spread in the context of asset pricing. This model abandons the one price universe, and by combining theories of No-Arbitrage and Expected Utility it proposes two price framework, one for buying and one for selling.

Chapter 3 gives an extended view onto Conic Finance Theory, by introducing main definitions and theorems. Afterward, distortion functions are introduced with an overview about several main functions used in the literature.

Chapter 4 starts by providing an introduction to the properties of a Brownian Motion and, specifically, a Geometric Brownian Motion. It continues with the derivation of the bid and ask prices of the European call and put options. After the derivation of the relevant formulas for the options' bid and ask prices, the chapter moves on to the description of the financial data used to obtain the presented results. It continues to present a range of combined events that explains the increases in the illiquidity premium during the course of the option maturity.

Chapter 5 sets out an extended model used for estimation of the effective bid and ask prices as well as the illiquidity premium. The chapter starts by the introduction of jump models and continues with the positive and negative attributes of these models. The chapter then proceeds with the introduction of Kou model. Afterwards, we calculate the distribution function of the stock price process that follows the Kou model. During the calculation steps, we use Residue Theorem from Complex Analysis and Inversion Formula from the Theory of Inverse Problems for densities.

Chapter 6 focuses on numerical estimations of the Kou model parameters. The chapter begins with the theoretical introduction of several models used in numerical calculation and moves to the evaluation of the parameters of the Kou model. After verification of the significance of the parameters by using statistical tests, the chapter finishes with the illiquidity premium calculation by using the bid and ask prices of European options and with a corresponding optimization technique.

Chapter 7 presents a new Early-Warning signaling for financial bubbles by benefiting from the theory of Conic Finance, from optimization and selected numerical methods. The chapter starts with the most famous historical bubbles and it continues by identifying how the illiquidity premium can be regarded as a main indicator for any upcoming financial trouble or bubble. For this case, we consider the U.S. markets since bubbles, mostly, have occurred in those markets.

The thesis concludes with Chap. 8 which presents a summary of the research undertaken and provides directions for future work in this area. It draws together our conclusions about the developed theory, the new findings and suggestions' and probable future enhancement of current research.

Reference

1. P. Shen, R.M. Starr, Market-makers' supply and pricing of financial market liquidity. Econ. Lett. **76**(1), 53–58 (2002)

Chapter 2
Review on Research Conducted

There are two main approaches in explaining the theory of the spread between bid and ask prices. The first one describes the role of inventory and its impact on prices formed by dealers. This approach focuses on the market maker's inventory problem, where buyers and sellers are liquidity traders without any informational advantages. The second one describes the role of information asymmetry by analyzing the role of trading insider information. In this approach, the market maker and liquidity traders are both uninformed regarding to the information-motivated traders. In this chapter, the notions *spread* and *bid-ask spread* will be used interchangeably.

2.1 Inventory Models

One of the earliest approaches in the price formation in market micro-structure theory is considered to be inventory models. In these models, equilibrium prices are the result of a dealer's obligation to offer liquidity, and face uncertainty in order flow. Basically saying dealer may end up with a position in which a price risk will arise. Hence, the spread in this case is assumed to be a compensation for bearing that risk.

Among the first scholars to propose spread as the liquidity measure was Demsetz [7] in his article dated back to 1968. Demsetz shows that spread can be regarded as the cost of immediacy or trading costs. This means that, spread would occur as the consequence of the time measure of the trade action since traders who are eager to do business at the earliest convenience have to pay a cost for this trade immediacy. The importance of limit orders as a source of liquidity has also been emphasized in the same article. The results of that paper show that the greater the activity (e.g., the number of trades) in the given stock the lower the spread, which indicates that the cost of non-synchronization between demand and supply for similar assets is a relation of the rate at which purchase and sell agreements reach the trading floor.

A. Karimov, *Identifying Stock Market Bubbles*, Contributions to Management Science, https://doi.org/10.1007/978-3-319-65009-8_2

Another scholarly article was written by Garman [9]. In his article, Garman argues that the ambiguity in the order flow is the main cause of uncertainty in the market makers' cash position, because buy orders will cost money while sell orders will generate cash. The article further indicates that the spread is an instrument that market maker employs to control the order flow, the corresponding cash position and the arising inventory level. More specifically, endogeneity of the probability of buying and selling orders depends on the listed bid and ask prices. Essentially, the article argues that higher bid prices will increase the probability of facing a trader who is selling while lower ask prices increase the probability of facing a trader who is buying. A change in the spread enables the market maker to control the inventory, but this in turn changes the cash position, which means that a trade-off is established between controlling inventory and the cash position.

Ahimud and Mendelson in [1] analyze a price-setting market maker in an economy with stochastic demand and supply where prices follow Poisson processes. In this analysis, the spreads depend on the stock inventory position of the market maker. Authors further derive the behavior of the spreads for demand and supply functions with linear nature and prove that on a transaction bases price behavior is inter-temporally dependent. In addition, the article proves that profits will disappear on a price dependence in cases where trade will be implemented contrary to the market-maker, which indicates that price-changes with serial dependency are in line with the market efficiency hypothesis.

Some inventory models deal with the optimization problems of the risk averse market maker. The market maker is simply a liquidity provider who maximizes the risk-return profile of the held portfolio, hence, yielding minimal inventory costs. In this articles, the main function of prices is adjustment of the stock position to the necessary level (i.e. to the level that maximizes market makers utility of expected wealth), therefore bid-ask spread is formed as a follow-up for this risk aversion.

In the context of Stoll [19], the spread can be seen as the cost of the vulnerability to distinct set of risks, like information, holding and order. The analyses are done for the investor or dealer willing to deviate from an optimal mean-variance efficient portfolio in order to facilitate trading. The spread therefore can be regarded as a compensation for the dis-utility that arises from this deviation. Additionally, these order costs consist of fixed-order charges that decline with the order size and information charges emerging as a result of the adverse selection from traders with better informed dealers. As in Garman [9], Stoll's model also treats order flow as a random process, controlled by the bid-ask prices. However, Stoll is more concerned with the costs of offering immediacy while Garman focuses on the equilibrium price under a random arrival of trades.

The model developed by Ho [11] is the multiperiod extension of Stoll [19]. It considers additional order uncertainty and indicates that the spread can be split into a risk premium and a risk neutral component. Specifically, holding costs rely upon the value of the order, the variance of the return, the risk-aversion and the wealth of the dealer, etc. The article assumes that when only holding costs are considered, the stock position on its own does not affect the trade volume; only the placement of the spread affects the trade volume. Moreover, the spread remains independent of the inventory level, and also changes with the market maker's planning horizon. In the models discussed by Stoll [19] and Ho [11], the market makers only, discriminate between different types of traders based on the volume.

One additional assumption of inventory models is that order flows are uncorrelated with future price movements. This is very unlikely as order flows contain information about the markets' perception about fundamentals. In this concept, *information models* can be seen as earliest models that incorporate distinct kinds of traders and the role of information onto the future price developments.

2.2 Information Models

The class of models that provide important insights of the role of adverse selection into the price formation process is called *information models*. They can be divided into two classes: one analyzes the impact of the traders with information on market makers and the other one uses statistical models to analyze the bid-ask spread.

2.2.1 *Informed Traders vs. Market Makers*

A first series of information-based models is introduced as *competitive-behavior models*. The main assumption of these models is that the market participants act competitively. Within this category we can include the works of Copeland and Galai [6], Easley and O'Hara [8] and Glosten and Milgrom [10].

Copeland and Galai [6] use a 2-period structure to analyze the markets with informed traders. In this structure, the interest of informed traders is obvious while the uninformed traders do business for exogenous reasons. Dealers unable to differentiate between these two types of investors, will apply positive spreads to make up for the expected loss that dealer will bring upon himself in case there is a probability of some investors being informed. The article also suggests that limit orders or quotes are disclosed to the prospect of being "*picked off*" when the market pricing varies, leading to unprofitable execution. Additionally, the asymmetry in the timing of the moves is at the origin of the adverse selection.

Unlike Copeland and Galai [6], Glosten and Milgrom [10] form an exchange with three participants: market maker, informed traders, and uninformed traders. All participants ought to be risk neutral and competitive. The article shows that the unbalanced information alone requires a positive spread, which relies upon the nature of the underlying information, the number of informed traders and the uninformed supply and demand elasticities. Prices form martingales and mirror all publicly available information, hence, the market is semi-strong form of efficiency. Additionally, adverse information costs bring almost no serial correlation in prices. Eventually, when the adverse selection is exceptionally inflated, spreads increase so much that markets close.

Both articles that we discussed above indicate that the likelihood of trade based on information can cause a deviation between bid and ask prices. Easley and O'Hara [8], on the contrary, show that the possibility of the same trade need not always leads to a spread. Depending on the market conditions, traders with information may determine to trade only considerable quantities, which will have no effect on the prices of small trades. The article also establishes that prices and spreads will vary across different trade volumes.

2.2.2 *Bid-Ask Spread as the Statistical Model*

There are numerous spread definitions. Some focus on the gap between the bid and ask prices and some like Roll-Indicator introduced by Roll [18] determine the adequate spread with regard to the first-order serial covariance of changes of prices. This indicator uses readily available time series of market prices, which makes it efficient and easy to use. In the same article, Roll considers an informationally competent market with zero trading costs, meaning that any price changes would be an explicit result of the release of new information. The paper presumes that in an efficient market, the likelihood of a trade occurring at the bid price is 0.5 and it is independent of past transactions. Roll argues that in such a market, with only an order processing component of the spread, the movement of transaction prices between bid and ask creates a *negative* first-order auto covariance of differences in transaction values. Using this relationship, he derives a plain estimator of the adequate spread:

$$S_{Roll} = 2\sqrt{-\mathrm{Cov}(\Delta P_t, \Delta P_{t-1})}. \tag{2.1}$$

The generalized version of the Roll model, introduced by Choi et al. [4], allows for the possibility of serial covariance in the chain of trade initiations. This means that the possibility of the new trade being initiated at the ask (bid) price given that the last trade occurred at the ask (bid) price may differ from 0.5. The authors concluded that the conditional probability of an extension could be bigger than 0.5 because large market orders often initiate trades with more than one participant as a counterparty. Hence, single trades will be recorded as multiple sequential trades in a ticker tape output, with ticker tape being the earliest digital electronic communications device, that transmitted stock price information over telegraph lines till 1970. The article also derives the following modified Roll estimator:

$$S_{CSS} = \frac{\sqrt{-\mathrm{Cov}(\Delta P_t, \Delta P_{t-1})}}{\pi}, \tag{2.2}$$

with π indicating the likelihood of a trade reversal and, moreover the first-order auto covariance of transaction prices being negative.

The costs of inventory holding, order processing, and adverse selection are three factors of the bid-ask spread analyzed by Stoll [20]. In this context, the relationship between abnormal trading volume and elements beside adverse selection costs can also force the negative relationship. This work also expresses that in the model, possibility of a trade switch regarding inventory control (a sell followed by a buy order or vice versa) is higher than $\frac{1}{2}$.

Huang and Stoll [13] used trade indicator models and generalized the methodology of Stoll [20] so that they would became able to assess the elements of the spread. In their article, spread has been decomposed into inventory, order processing and adverse selection components. Such a model assumes that all trades are of unitary size, and the "*true*" public information price of a stock evolves according to the following formula:

$$\begin{aligned} V_t =& V_{t-1} + \alpha \frac{S}{2} \text{ (unexpected change in inventory)} \\ &+ \text{(public information innovation)} \\ =& V_{t-1} + \alpha \frac{S}{2} \left(I_{t-1} - E\left[I_{t-1} \mid I_{t-2} \right] \right) + \varepsilon_t, \end{aligned} \tag{2.3}$$

where ε_t is an indicator for the public information innovation and I_t is an indicator function with the following specifications:

$$I_t = \begin{cases} 1, & \text{buyer initiated trade at time } t, \\ -1, & \text{seller initiated trade at time } t. \end{cases} \tag{2.4}$$

Hence, the expected change in inventory is simply:

$$E\left[I_{t-1} \mid I_{t-2} \right] = (1 - 2\pi) I_{t-2}, \tag{2.5}$$

with π being the reversal probability.

Lesmond in his article [15] allowed for the assessment of transaction costs by using the zero-returns measure, which Roll [18] had assumed to be non-real since transaction costs are not always accessible. Lesmond was also able to internally calculate the efficient transaction costs using historical data of security returns on a daily basis. This estimate essentially contains trading costs in order to capture the explicit cost of trade (i.e., the change between cost portions of sale and purchase).

Holden's [12] extended Roll's [18] assumption on securities being traded half of the time at the ask and the other half at the bid by introducing a measure which contains the probability that the stated end price may be the average of these prices. Holder gives two justifications for this. Firstly, the last trade of the day may be improving prices for both orders as a result of the buy and sell orders crossing the midpoint. Secondly, the trade price may be set at the average of closing bid and ask prices as the result of absence of sufficient trades on the particular day.

2.2.3 Introduction of Transaction Costs

Another approach, different from the information and inventory models, given in the literature is related with the transaction costs. The spread is assumed to be equal to the fee expenses for trading and could be associated with numerous experimental features of the asset under analysis, such as the flow of order and quantity of trade. Research also discusses liquid markets trading fees.

Constantinides [5] does not account for transaction costs in asset pricing, as authors research shows that the risk premium due to these costs is very small. He presumes a rather long holding period, and disputes that investors decrease the frequency and amount of trades when transaction costs are large. The same line of thinking has been used to reach an equilibrium condition with a definition of the illiquidity premium. The decrease in expected return on an illiquid asset makes the investor indifferent between the illiquid asset and a liquid asset.

By using martingales, Jouini and Kallal [14] investigated the no-arbitrage problem in securities markets under transaction costs and constraints on short sales. They showed that the lack of arbitrage is equivalent to the presence of a frictionless arbitrage-free process between bid and ask prices. Hence, all the constant proportional transaction cost models tested on frictionless arbitrage-free price process, are arbitrage-free. The converse is not always true.

Lo et al. [16] analyzed the effect of fixed transaction cost when investors have a continuous trading need. Investors receive a non-traded income continuously over time so that in the frictionless market, they trade continuously to share risks. However, the existence of transaction costs prevents investors from the continuous risk-sharing trading. The authors found that even a small transaction cost can lead to a large non-trading zone and generate a significant liquidity premium in asset prices. In the presence of transaction cost, the trading volume is finite. The increase of transaction cost, however, has marginal effects only on the trading volume.

2.3 Conic Finance

The models mentioned in Sects. 2.1 and 2.2 are based on empirical studies in relatively liquid markets. The spread between bid and ask prices in those markets have been justified by costs like the volume of trades and commissions. In this thesis, we will analyze derivative markets which are often much less liquid and requires market maker to have additional compensation for holding unhedgeable risks. Contrary to the models formed on the law of one price, the model proposed by Madan and Cherney [17] introduce two different prices for the price processes in the market: one price for buying and another price for selling. They simply abandon the law of one price and the no-arbitrage theory, and propose different formulas for bid and ask prices applicable to financial derivatives.

Carr et al. [2] introduce a new approach that combines the theories of arbitrage pricing and expected utility maximization. More importantly, the set of opportunities is expanded from arbitrage opportunities to a set of acceptable opportunities, and acceptable opportunities are defined as opportunities that a wide range of risk-averse investors would be willing to accept.

In a market with a vast amount of investment opportunities it is crucial to evaluate the performance of all different opportunities in order to decide which investment to make. The theory of Conic Finance is a new measure for performance and uses indices of acceptability to choose between different investment opportunities. The indices of acceptability are a mixture of the theory of expected utility maximization and the arbitrage pricing theory. Need for little information is an advantage of arbitrage pricing theory, but lack of strong implications to decide between two cash flows which are not arbitrage opportunities, when one cash flow might be more attractive to the investor than the other, is the major disadvantage of this theory. On the other hand, the theory of expected utility maximization is a powerful tool for the decision on the investment opportunity. This theory only works as long as the behavior of an investor is consistent with the *Von Neumann Morgenstern axioms* [21]. In practice this does not always hold and, more importantly, there are difficulties in correctly specifying the required inputs for optimization. In summary, the theory of expected utility maximization turns out not to be a good way to decide whether an investor should accept or reject an investment opportunity.

Cherny and Madan [3] presented a set of axioms that might be satisfied by measures of performance. They consider eight axioms: *quasi-concavity*, *monotonicity*, *scale invariance*, *Fatou property*, *law invariance*, *second-order stochastic dominance*, *arbitrage* and *expectation* consistency. When combined, the first four axioms define new acceptability indices. The other axioms are used to make additional comparisons. Cherny and Madan used performance measures that assign a value $\alpha(X)$ to a stochastic investment opportunity. A higher value of $\alpha(X)$ means a higher level of acceptability. These values enable financial managers to decide between different investment opportunities.

References

1. Y. Amihud, H. Mendelson, Dealership market: market-making with inventory. J. Finance Econ. **8**(1), 31–53 (1980)
2. P. Carr, H. Geman, D.B. Madan, Pricing and hedging in incomplete markets. J. Finance Econ. **62**(1), 131–167 (2001)
3. A. Cherny, D. Madan, New measures for performance evaluation. Rev. Financ. Stud. **22**(7), 2571–2606 (2009)
4. J.Y. Choi, D. Salandro, K. Shastri, On the estimation of bid-ask spreads: theory and evidence. J. Financ. Quant. Anal. **23**(02), 219–230 (1988)
5. G.M. Constantinides, Capital market equilibrium with transaction costs. J. Polit. Econ. **94**, 842–862 (1986)
6. T.E. Copeland, D. Galai, Information effects on the bid-ask spread. J. Finance **38**(5), 1457–1469 (1983)

7. H. Demsetz, The cost of transacting. Q. J. Econ. **82**(1), 33–53 (1968)
8. D. Easley, M. O'hara, Price, trade size, and information in securities markets. J. Financ. Econ. **19**(1), 69–90 (1987)
9. M.B. Garman, Market microstructure. J. Financ. Econ. **3**(3), 257–275 (1976)
10. L.R. Glosten, P.R. Milgrom, Bid, ask and transaction prices in a specialist market with heterogeneously informed traders. J. Financ. Econ. **14**(1), 71–100 (1985)
11. T. Ho, H.R. Stoll, Optimal dealer pricing under transactions and return uncertainty. J. Financ. Econ. **9**(1), 47–73 (1981)
12. C. Holden, New low-frequency liquidity measures. J. Financ. Markets **12**(4), 778–813 (2009)
13. R.D. Huang, H.R. Stoll, The components of the bid-ask spread: a general approach. Rev. Financ. Stud. **10**(4), 995–1034 (1997)
14. E. Jouini, H. Kallal, Martingales and arbitrage in securities markets with transaction costs. J. Econ. Theory **66**(1), 178–197 (1995)
15. D.A. Lesmond, J.P. Ogden, C.A. Trzcinka, A new estimate of transaction costs. Rev. Financ. Stud. **12**(5), 1113–1141 (1999)
16. A.W. Lo, H. Mamaysky, J. Wang, Asset prices and trading volume under fixed transactions costs, Technical Report, National Bureau of Economic Research (2001)
17. D.B. Madan, A. Cherny, Markets as a counterparty: an introduction to conic finance. Int. J. Theor. Appl. Finance **13**(08), 1149–1177 (2010)
18. R. Roll, A simple implicit measure of the effective bid-ask spread in an efficient market. J. Finance **39**(4), 1127–1139 (1984)
19. H.R. Stoll, The supply of dealer services in securities markets. J. Finance **33**(4), 1133–1151 (1978)
20. H.R. Stoll, Inferring the components of the bid-ask spread: theory and empirical tests. J. Finance **44**(1), 115–134 (1989)
21. H.R. Varian, The nonparametric approach to production analysis. Econometrica **52**, 579–597 (1984)

Chapter 3
Theory of Conic Finance

Some models try to determine whether an investor should invest in an uncertain opportunity or not, for example it can be done by specifying an index that measures the performance of a stochastic cash-flow on the basis of some specific characteristics. This reduces an investment decision to the comparison of a value attached to the cash-flow with the one that is attached to other cash-flows or with a certain benchmark. The main point here is to define which characteristics of stochastic cash-flows must be considered in order to develop an index that leads to sound investment decisions.

The field of Conic Finance specifies a new index of acceptability as improvement of the economic characteristics of the traditional performance measures. Theory integrates the full probability distribution of a zero costs stochastic cash-flow into the index, which is used to calculate a stress level that the stochastic cash-flow can tolerate, while being attractive for a wide range of investors. A higher level of maximum stress implies a higher level of acceptability that is attached to the stochastic cash-flow.

3.1 Conic Finance

In finance, transactions are based of buying or selling of a contracts or an opportunity with a stochastic cash-flow at a certain time in the future. Suppose that a random variable X, a zero cost stochastic cash-flow at some particular time, is identified on probability space $(\Omega, \mathcal{F}, \mathbb{P})$. The assumption of a zero cost investment is not entirely realistic, but it does not affect the generality of the theory, as the premium, cost of the opportunity, can be borrowed at a risk-free rate and paid back at the final payoff date of the cash-flow. The time value of money is important, that is why suitable discount factors should be used in calculations.

A. Karimov, *Identifying Stock Market Bubbles*, Contributions to Management Science, https://doi.org/10.1007/978-3-319-65009-8_3

It is crucial to evaluate the performance of each investment when there are numerous opportunities in the market in order to decide which investment to make. In this regards, Conic Finance as a theory is based on indices of acceptability as a new measure for performance. These indices of acceptability are a combination of the *Theory of Expected Utility Maximization* and the *Arbitrage Pricing Theory*. Arbitrage pricing theory only needs little information, which is an advantage, but it cannot decide between two cash flows that are not arbitrage opportunities, although one cash flow might be more attractive to the investor, which is a disadvantage. The theory of Expected Utility Maximization is a very effective tool to decide which investment opportunity should be accepted and which not. It works as long as the behavior of an investor is guided by the *Von Neumann Morgenstern* axioms [5]. Unfortunately this is not always the case in practice and, even more importantly, there are difficulties in correctly specifying the required inputs for optimization. In brief, the theory of expected utility maximization turns out not to be a good way to decide whether an investor should accept or reject an investment opportunity.

Carr et al. [2] introduced a new approach that combines the theories of arbitrage pricing and expected utility maximization. Importance of this concept is the idea that the set of opportunities is expanded from arbitrage opportunities to a set of acceptable opportunities. Acceptable opportunities are defined as opportunities that a wide range of risk-averse investors will be willing to accept.

With a view to escape the details related to the finite moments, the subgroup of delimited random variables defined by $L^{\infty} = L^{\infty}(\Omega, \mathcal{F}, \mathbb{P})$ will be used in this thesis and, moreover a map of $\alpha : L^{\infty} \rightarrow \overline{\mathbb{R}}^{+}$, where $\overline{\mathbb{R}}^{+} := [0, \infty]$, will be defined as the measure of performance. For a random variable $X \in L^{\infty}$ which could be a cash flow from a trading strategy, $\alpha(X)$ will be measuring the performance or quality of X.

3.2 Conic Finance in Practice

The space of traded cash flows, which are all bounded random variables, is assumed to be a base probability space $(\Omega, \mathcal{F}, \mathbb{P})$. Moreover, there is a unique pricing measure $\mathbb{Q}$, equivalent to $\mathbb{P}$, and a complete market. In these regards, we assume that we have a cash flow, CF, that has a price of p such that $p \neq 0$; then the difference $X = CF - p$ can be formed with zero price and any quantity of the cash flow X may be traded with the market. Moreover, the set of cash flows traded in liquid markets will have the property $E^{\mathbb{Q}}(X) \geq 0$.

Specifically, if a cash flow CF has a price of p and the cash flow of $X = CF - p$ is bought by the market, then we may say that in an incomplete market the price of CF, indicated by p, will be a bid price. It would be impossible for the market to sell and keep $-X = p - CF$, as an individual will be expected to pay a spread s, to the market, which will be ready to get $(p + s) - CF$. In this setup, the set of cash-flows is no longer closed under negation due to the fact on incompleteness. The market model is identified by defining the set of cash flows which is the cash flows smaller than the one in the liquid market.

It must be noted that the set of marketed cash flows is convex, closed under scaling, and also only a linear mixture of any amount of these cash flows can be marketed. The generic model for the set sustains the given two characteristics and, thus, the set of cash flows would nevertheless be a convex cone. Moreover, since all non-negative cash flows are marketed, it is obvious that the set consisting of marketed cash flows could be a convex cone and this set will include the non-negative cash flows. It can be easily deducted from Artzner et al. [1] that this set will be made of all cash flows X satisfying the following condition:

$$E^{\mathbb{Q}}(X) \geq 0 \text{ for any } \mathbb{Q} \in D, \tag{3.1}$$

where D is a set of measures that are equivalent to $\mathbb{P}$, and also these sets are convex. It is supposed that the risk-neutral universe is also included in D.

To achieve this degree of change for the set of given marketed cash flows, Cherny and Madan [3] proposed an acceptability index that permits to deliver the cash flows which are acceptable or equivalently marketed at level γ.

An *index of acceptability* is a function α that links every delimited random variable X to a number $\alpha(X)$ in $[0, \infty]$, which is named as acceptability level of X. Such an α would have four features, as we have discussed in Sect. 3.1. First of all, if 2 distinct random variables are acceptable at a level γ, then so does the sum of these 2 random variables. Secondly, if one random variable is acceptable at a level γ and another one prevails over this random variable, then the second random variable will also be acceptable at the same level. Thirdly, for a random variable Y acceptable at a level γ and for any positive scalar of a, a random variable aY would also be acceptable at the same level. Lastly, if the set of random variables $(X_n)_{n\geq 1}$ are all acceptable at a level γ and $X_n \to X$ in probability then the random variable of X is also acceptable at the same level.

Cherny and Madan in their paper [3] provided examples of four new law-invariant indices of acceptability and, hence, *distortion functions* related with them. These indices, *aimin*, *aimax*, *aimaxmin*, and *aiminmax* (where "ai" means the "acceptability index") are established by taking into account 4 distinctive categories of sampling:

- *aimin*—a sample is formed as the expected value of the smallest of several selections from the cash flow distribution. The risk measure associated with this acceptability index is called as *minvar*.
- *aimax*—a sample is formed by making several selections from the distribution and taking the biggest to get the cash flow distribution. The relevant risk measure is called as *maxvar*.
- *aiminmax* and *aimaxmin*—merge the aforementioned processes in different consequences. The applicable risk measures are called *minmaxvar* and *maxminvar* appropriately.

An elementary link exists between the groups of probability measures and acceptability indices. Madan and Cherny in [4] state that any acceptability index is linked to an increasing one-parameter family of sets of probability measures so that the value $\alpha(X)$ is the largest level γ such that X has a positive expectation under each measure from the set corresponding to the level γ. In other words, the acceptability at level γ means that all measures from the corresponding family value X are positive. The index $\alpha(X)$ is then the highest level of acceptability attained.

For the acceptability index of α, and for each $\gamma > 0$ there is an absolutely continuous set of probability measures of D_γ equivalent to the initial probability measure of $\mathbb{P}$. A random variable is acceptability at level γ if and only if it has a positive expectation under each measure from D_γ. In regards to the family $(D_\gamma)_{\gamma\geq 0}$, the index α will satisfy the subsequent theorem (for the proof see [3]).

Theorem 3.1 ([3]) *A map α from the set of bounded random variables on a base probability space of $(\Omega, \mathcal{F}, \mathbb{P})$ to the extended half-line, $[0,\infty]$, is an acceptability index if and only if there exists a family of sets of probability measures $(\mathbb{D}_\gamma)_{\gamma\geq 0}$ equivalent to $\mathbb{P}$ such that $\mathbb{D}_{\gamma_1} \subseteq \mathbb{D}_{\gamma_2}$ for $\gamma_1 \leq \gamma_2$, and*

$$\alpha(X) = \sup\left\{\gamma \geq 0 : E^Q(X) \geq 0 \, for\ all\ Q \in \mathbb{D}_\gamma\right\} \tag{3.2}$$

where $\sup\emptyset = 0$.

Suppose that the trader sells a cash flow, X, for which driven by competition he charges a minimal price of a. Notwithstanding, the emerging remaining cash flow, $a - X$, ought to be α-acceptable at level γ. Hence, this price, a, would be the *ask price* of X. For $a-X$ to be acceptable at level γ, the price a must exceed $E^{\mathbb{Q}}(X) \geq 0$ for any $\mathbb{Q} \in D_\gamma$, so the minimal price would be derived using the following formula:

$$\begin{aligned} a_\gamma(X) &= \inf\{a : \alpha(a - X) \geq \gamma\} \\ &= \inf\{a : E^{\mathbb{Q}}[a - X] \geq \gamma \text{ for any } \mathbb{Q} \in D_\gamma\} = \sup_{\mathbb{Q}\in D_\gamma} E^{\mathbb{Q}}[X]. \end{aligned} \tag{3.3}$$

For the buy side, when the trader buys a cash flow, X, again driven by competition he charges a maximum price of b. Nevertheless, the emerging remaining cash flow, $X - b$, again ought to be acceptable at level γ. Hence, this price, b, would be the *bid price* and will have the following expression:

$$b_\gamma(X) = \inf_{\mathbb{Q}\in D_\gamma} E^{\mathbb{Q}}[X]\,. \tag{3.4}$$

Operational and parametric models for cash flows can be formulated by restricting the characteristics of acceptability so that these models would completely map the probability law of the imminent risk. Thus, in order to check the acceptability at a level γ for a random variable X, the only information needed is the CDF, F_X, of this process. Generally, investors and traders are caring about the relationship between the future and presently held risks, and the probability law on its own is not an adequate model for the market acceptability.

As proposed by Cherny and Madan in [3], parameter family of distortion functions can be used to formulate an operational index of acceptability. A *distortion function*, indicated by Ψ, is an increasing concave function from $[0, 1]$ onto $[0, 1]$; when applied to a cumulative distribution function, it is said to "*distort*" it at a rate specified by the parameter γ. Because of the concavity, the distortion functions will assign higher weights to lower outcomes of the random variable X, and higher outcomes will be weighted lighter. Hence, the larger γ, the lower becomes the distorted expectation of a cash flow X, which is characterized as:

$$\alpha\left(X\right) = \sup\left\{\gamma \geq 0 : \int_{-\infty}^{\infty} xd\Psi^{\gamma}\left(F_X\left(x\right)\right) \geq 0\right\}. \tag{3.5}$$

or

$$E^{\mathbb{Q}_\gamma}(X) = \int_{-\infty}^{\infty} xd\Psi^{\gamma}(F_X(x)), \tag{3.6}$$

where X is the random variable, F is the distribution function of X and $\gamma \geq 0$. The density $\Psi^{\gamma}(F_X(X))$ is identified in regard to the initial measure of $\mathbb{P}$. We have to indicate that for a concave distortion the new density gives more weight to the losses when $F_X(X)$ is near zero, and it decreases weights of the gains when $F_X(X)$ is near one.

The value $\alpha(X)$ expressed by Eq. (3.5) can be numerically computed if we have the distribution function of X.

3.3 Distortion Functions

The distortion function performs a crucial part in the realization of explicit bid and asks prices. As discussed, there are different distortion functions that can be used. Although some of them are not preferred, due to some undesirable properties, there are still a lot of different distortion functions to choose from. Therefore, an interesting question might be whether there is one particular distortion function that is best to use. Madan and Cherny [3] conclude that *maxminvar* and *minmaxvar* provide relatively similar results and prefer one of them to *maxvar* and *minvar*. The Wang transform that is used in this thesis is similar to *minmaxvar*. At this moment, there is no scientific proof that one particular distortion function is best to use, so further research on this topic is recommended.

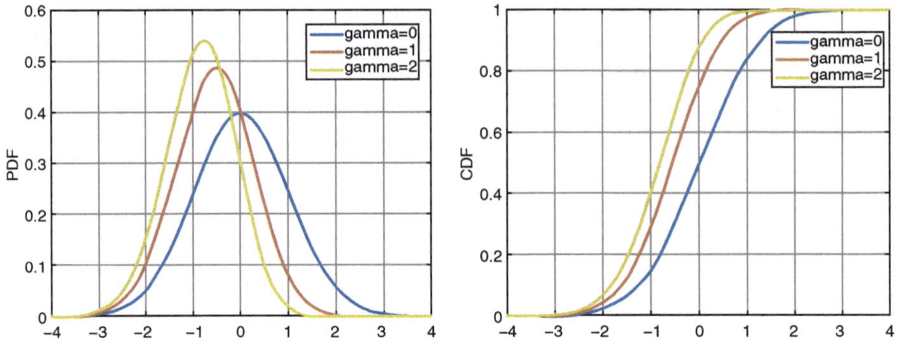

Fig. 3.1 *Minvar* distortion function on a standard normal distribution

3.3.1 *Minvar*

$$\Psi^{\gamma}(u) = 1 - (1-u)^{1+\gamma}, \quad u \in [0,1], \quad \gamma \geq 0. \tag{3.7}$$

This distortion function is based on the risk measure *minvar*, which corresponds to acceptability index *aimin*. A disadvantage of *minvar* is that the maximal weight on big losses is relatively small and hence this distortion function does not have enough relevance in economic theory. Consumers are generally risk-averse and absolutely do not want large losses, but this risk measure is too lenient towards large losses and, therefore, not risk-averse enough to be a good risk measure. *minvar* is rarely used, because there are more relevant alternatives. Figure 3.1 shows the effects of the *minvar* distortion on the standard normal distribution. The PDF and CDF shift to the left.

3.3.2 *Maxvar*

$$\Psi^{\gamma}(u) = u^{\frac{1}{1+\gamma}}, \quad u \in [0,1], \quad \gamma \geq 0. \tag{3.8}$$

This distortion function is based on risk measure *maxvar*, which corresponds to acceptability index *aimax*. *Maxvar* is more promising than the previous risk measure, but also has a potential drawback. It does not discount large gains and, therefore, is rarely used. The effects of this distortion is depicted in Fig. 3.2.

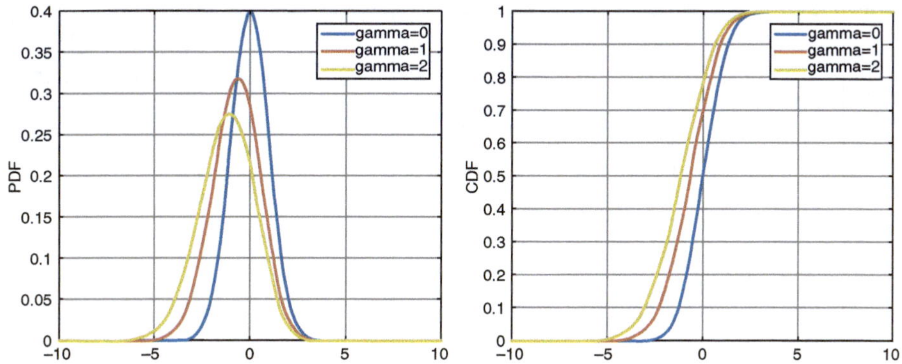

Fig. 3.2 *Maxvar* distortion function on a standard normal distribution

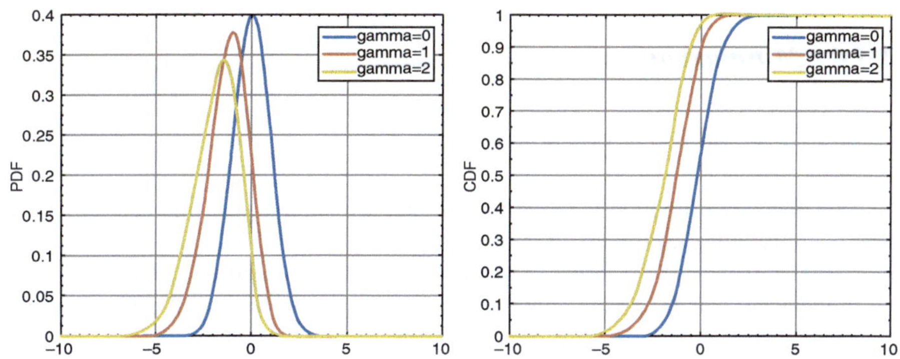

Fig. 3.3 *Maxminvar* distortion function on a standard normal distribution

3.3.3 *Maxminvar*

$$\Psi^{\gamma}(u) = \left(1 - (1-u)^{1+\gamma}\right)^{\frac{1}{1+\gamma}}, \quad u \in [0,1], \quad \gamma \geq 0. \tag{3.9}$$

This distortion function is based on risk measure *maxminvar* and the corresponding acceptability index *aimaxmin*. It is obtained by first using a *minvar* procedure and then a *maxvar* procedure. This risk measure does not have the drawbacks of the previous two and satisfies all eight desirable axioms. See Fig. 3.3 for the effects of the application of this distortion function.

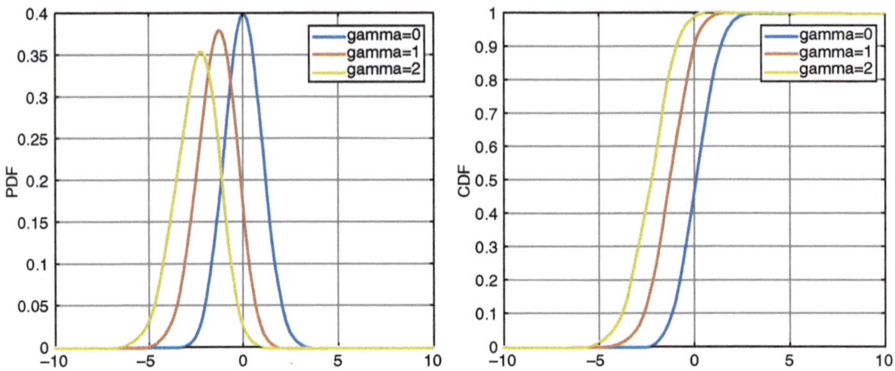

Fig. 3.4 *Minmaxvar* distortion function on a standard normal distribution

3.3.4 *Minmaxvar*

$$\Psi^{\gamma}(u) = 1 - \left(1 - u^{\frac{1}{1+\gamma}}\right)^{1+\gamma}, \quad u \in [0,1], \quad \gamma \geq 0. \tag{3.10}$$

This distortion function is based on risk measure *minmaxvar* and corresponding acceptability index *aiminmax*. It is obtained by first using a *maxvar* procedure and then a *minvar* procedure. *Minmaxvar* satisfies all eight desirable axioms and has no drawbacks. It is relatively comparable to maxminvar, but small differences are noticeable when comparing Fig. 3.3 to Fig. 3.4.

Apart from the four distortion functions introduced by Madan and Cherny [3], there are also several other distortion functions that are used in risk management. The Wang transform [6] is an important one.

3.3.5 *Wang Transform*

$$\Psi^{\gamma}(u) = \Phi(\Phi^{-1}(u) + \gamma), \quad u \in [0,1], \quad \gamma \geq 0. \tag{3.11}$$

This distortion function is introduced by Wang and shifts the original distribution. These effects are visible in Fig. 3.5. Here, Φ is the standard normal CDF.

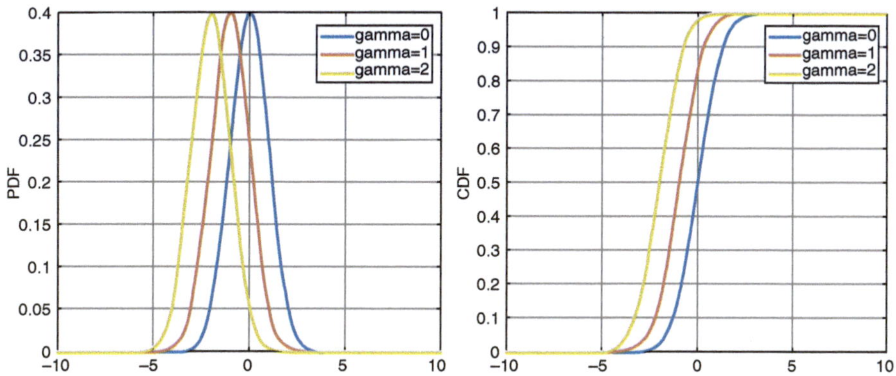

Fig. 3.5 Wang transform on a standard normal distribution

References

1. P. Artzner, F. Delbaen, J.-M. Eber, D. Heath, Coherent measures of risk. Math. Fin. **9**(3), 203–228 (1999)
2. P. Carr, H. Geman, D.B. Madan, Pricing and hedging in incomplete markets. J. Financ. Econ. **62**(1), 131–167 (2001)
3. A. Cherny, D. Madan, New measures for performance evaluation. Rev. Financ. Stud. **22**(7), 2571–2606 (2009)
4. D.B. Madan, A. Cherny, Markets as a counterparty: an introduction to conic finance. Int. J. Theor. Appl. Finance **13**(08), 1149–1177 (2010)
5. H.R. Varian, The nonparametric approach to production analysis. Econometrica **52**, 579–597 (1984)
6. S.S. Wang, A universal framework for pricing financial and insurance risks. Astin Bull. **32**(2), 213–234 (2002)

Chapter 4
Stock Prices Follow a Brownian Motion

In both expressions for calculating prices of put and call options the outcome rely on the CDF of underlying process. In this instance, distribution function of the stock process is needed to calibrate gamma. This part of the thesis assumes that stock or index price processes are lognormally distributed. We will start this chapter by introducing a Geometric Brownian Motion and then we will move onto the calculation of the illiquidity premium. Moreover, by employing the assumptions of Black-Scholes model, we will infer the distribution function of stock price process.

4.1 Geometric Brownian Motion: Introduction

Brownian motion stands in one of the central roles in probability theory, theory of stochastic processes, and also in finance. We will start this section of the thesis with a description of this process, and will move on to listing of several of the basic properties of Brownian motion.

Definition 4.1 ([8]) On a probability space $(\Omega, \mathcal{F}, \mathbb{P})$, a real-valued stochastic process $(W_t)_{t\geq 0}$ is called a *Brownian motion* or a *Wiener process*, if it satisfies the following conditions:

- $(W_t)_{t\geq 0}$ has *independent increments*. In other words, $W_y - W_t$ and $W_s - W_v$ are independent for $v < s \leq t < y$;
- $(W_t)_{t\geq 0}$ has *continuous paths*;
- $W_t - W_s$ follows a Normal Distribution function with zero mean and variance of $t - s$ or $W_t - W_s \sim N(0, t - s)$ for $0 \leq s < t$.

The Brownian motion is called *standard* if it starts at 0, i.e., $\mathbb{P}(W_0 = 0) = 1$.

The finite-dimensional distributions of Brownian motion are multivariate Gaussian, so, $(W_t)_{t\geq 0}$ is a Gaussian process. From the definition, we know that $W_t - W_s$

A. Karimov, *Identifying Stock Market Bubbles*, Contributions to Management Science, https://doi.org/10.1007/978-3-319-65009-8_4

will have the same distribution as $W_{t-s} - W_0 = W_{t-s}$, which is an $N(0, t-s)$ distribution.

From the above statement we can deduct that a Brownian motion has expectation operator with the value of,

$$\mathbb{E}(W_t) = 0 \quad (t \geq 0).$$

It has a covariance value of

$$Cov(W_t, W_s) = \min\{s, t\} \quad (s, t \geq 0).$$

Lévy's characterization is often considered as another way to define and identify Brownian motions.

Theorem 4.1 (Lévy's Characterization [8]) *Let $(\Omega, \mathcal{F}, (\mathcal{F}_t)_{t\geq 0}, \mathbb{P})$ be a filtered probability space. Let $(W_t)_{t\geq 0}$ be an adapted continuous local martingale in regards to $(\Omega, \mathcal{F}, (\mathcal{F}_t)_{t\geq 0}, \mathbb{P})$. In this case, W is a Brownian motion if and only if the quadratic variation of W is equal to t, i.e., $< W, W >_t = t$.*

Brownian motion is assumed to be in the nature of the stock markets, the foreign exchange markets, commodity markets and bond markets. In these markets assets are changing within very small time and position intervals which happens continually, and this is in the very characteristics of the Brownian motion. Essentially, all financial asset pricing and derivatives pricing models are based on the mathematical algorithms describing Brownian motions. These mathematical models are of main importance to the work that is being done on market models and risk analysis. One of these models is the Geometric Brownian Motion which has the following definition.

Definition 4.2 ([10]) A stochastic process $(S_t)_{t\geq 0}$ on a probability space of $(\Omega, \mathcal{F}, \mathbb{P})$ is said to follow a *Geometric Brownian Motion* if it satisfies the stochastic differential equation

$$dS_t = S_t(\mu dt + \sigma dW_t), \tag{4.1}$$

with μ being a drift term and σ being a volatility.

Equation (4.1) has a below given solution:

$$S_t = S_0 e^{(\mu - \frac{\sigma^2}{2})t + \sigma W_t} \quad (t \geq 0), \tag{4.2}$$

where S_0 represents the initial value of stochastic process. The same model is also used in pricing of option contracts.

Definition 4.3 ([3]) An *option* is the contract between two parties in which one party has the right but not the obligation to buy or sell an underlying asset, subject to the contract.

The rights beyond obligations given under the option contract has a financial value, so option buyers must pay a fee in order to purchase these rights, which basically makes options an asset to buyers. The given assets infer or derive their prices from their underlying assets, that is why they are called derivative assets or derivative contract. Nowadays, option pricing techniques, by using stochastic calculus, are among the most mathematically complex areas of finance [3].

4.2 Option Pricing with Geometric Brownian Motion

Knowing acceptability indices and distortion functions provides an opportunity to compute expressions for bid and ask prices. It is critical to notice that in Conic Finance the market is seen as a counterparty, willing to accept all stochastic process or cash-flows X that have an acceptability level of at least γ. The ask price $a_\gamma(X)$ is what the market asks from the trader, so the trader needs to pay that price to buy it. The bid price $b_\gamma(X)$ is the price that a trader gets for selling an asset to the market. The market makers will set their price in such a manner that the payoff of variable X is acceptable to them at level γ: specifically, for the case of an ask price the market maker will require a random variable of $a - X$ to be acceptable at the level of γ, which can be given in the inequality form as $\alpha(a - X) \geq \gamma$, and for the case of a bid price market maker will be satisfied in case when $X - b$ is acceptable at the level of γ, or $\alpha(X - b) \geq \gamma$. In the formulas below, with the use of Riemann-Stieltjes integral in (3.6), we will derive the ask price and, then, the bid price of the financial instruments:

$$\alpha(a - X) \geq \gamma \Longleftrightarrow \int_{-\infty}^{\infty} x d\Psi^{\gamma}(F_{a-X}(x)) \geq 0 \Longleftrightarrow a + \int_{-\infty}^{\infty} x d\Psi^{\gamma}(F_{-X}(x)) \geq 0,$$

$$a_\gamma(X) = -\int_{-\infty}^{\infty} x d\Psi^{\gamma}(F_{-X}(x)). \tag{4.3}$$

Analogously, for $b_\gamma(X)$, the bid price, we obtain:

$$\alpha(X - b) \geq \gamma \Longleftrightarrow \int_{-\infty}^{\infty} x d\Psi^{\gamma}(F_{X-b}(x)) \geq 0 \Longleftrightarrow -b + \int_{-\infty}^{\infty} x d\Psi^{\gamma}(F_{X}(x)) \geq 0,$$

$$b_\gamma(X) = \int_{-\infty}^{\infty} x d\Psi^{\gamma}(F_{X}(x)). \tag{4.4}$$

Theoretically, the ask price must be more than the bid price, because of the concave characteristics of the distortion functions. The parameter γ, fitting the theoretical bid-ask prices around the historical market prices will be called, the *Implied Illiquidity Parameter* or the *Illiquidity Premium* from here afterward.

It is assumed that $(S_t)_{t\geq 0}$, the stock prices at times $t \geq 0$, follow a Geometric Brownian Motion [8], with a stochastic differential equation having the following expression:

$$dS_t = \mu S_t dt + \sigma S_t dW_t \quad (t \geq 0), \tag{4.5}$$

where W_t $(t \geq 0)$ a Brownian Motion, reflects uncertainty of the price process in a standardized way.

The solution of Eq. (4.5) can be calculated to be equal to the following formula:

$$S_t = S_0 e^{Y_t}, \tag{4.6}$$

where $Y_t = \sigma W_t + (\mu - \frac{\sigma^2}{2})t$, $(t \geq 0)$ is a Brownian Motion with drift term μ and scaling or volatility term of σ, with respect to probability measure $\mathbb{P}$. Moreover, $S_0 > 0$ is assumed to be the initial value of price process. The Black-Scholes model will be used to derive the distribution function of S_t $(t \geq 0)$ [1].

With the help of Itô's Lemma which is shown in below, we can identify the distribution of the given stock price process. Consider a process X_t, for $t \geq t_0$, of the form:

$$X_t = X_{t_0} + \int_{t_0}^{t} b(u, X_u)du + \int_{t_0}^{t} \sigma(u, X_u)dW_u,$$

where $b(t, X_t)$ and $\sigma(t, X_t)$ are continuous nonanticipating processes with $E^Q \int_{t_0}^{t} \sigma(u, X_u)^2 du < \infty$, then the Itô's Lemma will have the following representation:

Lemma 4.1 ([4]) *For every twice differentiable function of $F(x, t)$, defined in $\mathcal{L}^2(\Omega \times [0, \infty])$, the Itô Formula is explained as follows:*

$$dF(x, t) = \frac{\partial F(x, t)}{\partial t} dt + \frac{\partial F(x, t)}{\partial x} dx + \frac{1}{2} \frac{\partial^2 F(x, t)}{\partial x^2} (dx)^2. \tag{4.7}$$

Let us calculate the above formula for the case of $F(S_t, t) = \ln S_t$, and recall that $dS_t = S_t(\mu dt + \sigma dW_t)$,

$$\begin{aligned} dF(S_t, t) &= \frac{\partial F(X_t, t)}{\partial t} dt + \frac{\partial F(X_t, t)}{\partial X_t} dX_t + \frac{1}{2} \frac{\partial^2 F(X_t, t)}{\partial X_t^2} (dX_t)^2 \\ &= \frac{\partial \ln S_t}{\partial S_t} dS_t + \frac{1}{2} \frac{\partial^2 \ln S_t}{\partial S_t^2} (dS_t)^2 = \frac{1}{S_t} dS_t - \frac{1}{2} \frac{1}{S_t^2} (dS_t)^2 \\ &= \frac{1}{S_t} S_t(\mu dt + \sigma dW_t) - \frac{1}{2} \frac{1}{S_t^2} S_t^2 \sigma^2 dt \\ &= \mu dt + \sigma dW_t - \frac{1}{2} \sigma^2 dt = (\mu - \frac{1}{2} \sigma^2) dt + \sigma dW_t. \end{aligned}$$

In the derivation of the above calculations we used the well-known fact that the quadratic variation of the standard Brownian motion simply is the time t [9].

If we go one step further in the above calculations, we will have the following expression of the stochastic differential equation:

$$d\ln S_t = (\mu - \frac{1}{2}\sigma^2)dt + \sigma dW_t.$$

From here we may simple deduct that the subsequent is true:

$$\ln S_t = \ln s_0 + (\mu - \frac{1}{2}\sigma^2)t + \sigma W_t, \tag{4.8}$$

or,

$$S_t = s_0 e^{(\mu - \frac{1}{2}\sigma^2)t + \sigma W_t}, \tag{4.9}$$

where s_0 denotes the initial value of the price process. From Eq. (4.8) we may conclude that the process $(\ln S_t)_{t\geq 0}$ has the Normal Distribution with mean $(\mu - \frac{1}{2}\sigma^2)t$ and variance of $\sigma^2 t$, which means that the process $(S_t)_{t\geq 0}$ has the Lognormal Distribution with the same parameters.

If we do the same calculation steps for the function of $F(S_T, T) = \ln S_T$, we will have the following results:

$$\ln S_T = \ln s_0 + (\mu - \frac{1}{2}\sigma^2)T + \sigma W_T, \tag{4.10}$$

or,

$$S_T = s_0 e^{(\mu - \frac{1}{2}\sigma^2)T + \sigma W_T}. \tag{4.11}$$

Let us assume that our price process given by Eq. (4.11) will start at time t instead of $t_0 = 0$, then Eq. (4.11) will have the following representation:

$$\begin{aligned} S_T &= s_0 e^{(\mu - \frac{1}{2}\sigma^2)T + \sigma W_T} = S_t e^{(\mu - \frac{1}{2}\sigma^2)(T-t) + \sigma(W_T - W_t)} \\ &= e^{(\mu - \frac{1}{2}\sigma^2)(T-t) + \ln S_t + \sigma(W_T - W_t)}; \end{aligned}$$

hence the following distribution function for the price process holds true:

$$S_T \overset{dist}{=} \text{lognormal}\left[\left(\mu - \frac{1}{2}\sigma^2\right)(T-t) + \ln S_t, \sigma^2 (T-t)\right]. \tag{4.12}$$

It is known from measure theory that under the risk-neutral measure $\mathbb{Q}$, the expected return of all securities is equal to the risk free rate, r, [5]; so if we change μ to r, we will get:

$$S_T \overset{dist}{=} \text{lognormal}\left[\left(r - \frac{1}{2}\sigma^2\right)(T-t) + \ln S_t, \sigma^2 (T-t)\right], \tag{4.13}$$

or

$$F_S(x) = \Phi\left[\frac{\ln\frac{x}{S_t} - \left(r - \frac{1}{2}\sigma^2\right)(T-t)}{\sigma\sqrt{T-t}}\right]. \tag{4.14}$$

with Φ being the standard normal cumulative distribution function.

After calculating the distribution function of the stock price process, we have to apply the distortion function to derive the bid and ask prices. When we apply the Wang transform of Ψ^{γ}, the distortion function, to the distribution function of F_{S_T}, we obtain another lognormal distribution function with mean $\mu^* := \mu + \gamma\sigma$, and variance $\sigma^* := \sigma$, and Ψ^{γ} will have the following representation:

$$\Psi^{\gamma}\left(F_{S_T}(x)\right) := \Phi\left[\frac{\ln\frac{x}{S_t} - \left(r - \frac{1}{2}\sigma^2\right)(T-t) + \gamma\sigma\sqrt{T-t}}{\sigma\sqrt{T-t}}\right]. \tag{4.15}$$

4.3 Bid-Ask Prices of European Options Under Brownian Motion

The price of the European call option can be expressed by the subsequent function:

$$C_T = \max\{(S_T - K), 0\}, \tag{4.16}$$

with S_t being the index price, K being the strike price, and T being the maturity time of the option.

By using Eq. (4.4) we will get the bid price of the call option:

$$\begin{aligned} b_{\gamma}(C) &= \int_0^{\infty} x\, d\Psi^{\gamma}(F_{C_T}(x)) \\ &= \int_K^{\infty} (x-K)\, d\Psi^{\gamma}(F_{S_T}(x)) \\ &= \int_K^{\infty} x\, d\Psi^{\gamma}(F_{S_T}(x)) - \int_K^{\infty} K\, d\Psi^{\gamma}(F_{S_T}(x)) =: A - B. \end{aligned} \tag{4.17}$$

By employing change of variables and basic calculus techniques we can derive the first integral in Eq. (4.17) as follows (for more detailed calculation please see Appendix B):

$$\begin{aligned} A &= \int_K^{\infty} x\, d\Psi^{\gamma}(F_{S_T}(x)) \\ &= e^{\ln S_t + r(T-t) - \gamma\sigma\sqrt{T-t}} \cdot \Phi\left[\frac{\ln S_t + (r + \frac{1}{2}\sigma^2)(T-t) - \gamma\sigma\sqrt{T-t} - \ln K}{\sigma\sqrt{T-t}}\right], \end{aligned}$$

and the second part of the above integral can be calculated as:

$$B = \int_K^\infty K \, d\Psi^\gamma(F_{S_T}(x)) = K\left(1 - \Phi\left[\frac{\ln\frac{K}{S_t} - \left(r - \frac{1}{2}\sigma^2\right)(T-t) + \gamma\sigma\sqrt{T-t}}{\sigma\sqrt{T-t}}\right]\right).$$

Black-Scholes model assumes that the starting premium is borrowed at the risk-free rate, therefore, we will have to use a suitable discount factor. Hence, the price needs to be multiplied by discount factor $e^{-r(T-t)}$, which leads to the following expression for the bid price:

$$b_\gamma(C) = S_t e^{-\gamma\sigma\sqrt{T-t}} \cdot \Phi\left[\frac{\ln S_t + (r + \frac{1}{2}\sigma^2)(T-t) - \gamma\sigma\sqrt{T-t} - \ln K}{\sigma\sqrt{T-t}}\right]$$
$$- e^{-r(T-t)} K\left(1 - \Phi\left[\frac{\ln\frac{K}{S_t} - \left(r - \frac{1}{2}\sigma^2\right)(T-t) + \gamma\sigma\sqrt{T-t}}{\sigma\sqrt{T-t}}\right]\right).$$

Similarly, by using Eq. (4.3) we will get the ask price of the European put option, where the option price process will be expressed by the following value:

$$P_T = \max\{(K - S_T), 0\}, \tag{4.18}$$

with S_T being the index price, K the strike price, and T the maturity time of the option. Employing Eq. (4.3) we will get the ask price of the put option as given subsequently:

$$\begin{aligned}
a_\gamma(P) &= -\int_{-\infty}^0 x \, d\Psi^\gamma(F_{-P_T}(x)) \\
&= -\int_{-\infty}^0 x \, d\Psi^\gamma(F_{S_T}(K + x)) \\
&= -\int_0^\infty x \, d\Psi^\gamma(F_{S_T}(K - x)) \\
&= \int_0^K (K - x) \, d\Psi^\gamma(F_{S_T}(x)) \\
&= \int_K^\infty (K - x) \, d\Phi\left[\frac{\ln\frac{K}{S_t} - \left(r - \frac{1}{2}\sigma^2\right)(T-t) + \gamma\sigma\sqrt{T-t}}{\sigma\sqrt{T-t}}\right].
\end{aligned} \tag{4.19}$$

By conducting the same calculation steps that we used during the calculation of the bid price of the call option, we will get the subsequent expression for ask price

Table 4.1 Prices of European call and put options

Option	Price	d_1	d_2
b_γ (C)	$S_t e^{-\gamma\sigma\sqrt{T-t}}\Phi(d_1) - e^{-r(T-t)}K\Phi(d_2)$	$\frac{\ln\frac{S_t}{K}+(r+\frac{1}{2}\sigma^2)(T-t)-\gamma\sigma\sqrt{T-t}}{\sigma\sqrt{T-t}}$	$d_1 - \sigma\sqrt{T-t}$
a_γ (C)	$S_t e^{\gamma\sigma\sqrt{T-t}}\Phi(d_1) - e^{-r(T-t)}K\Phi(d_2)$	$\frac{\ln\frac{S_t}{K}-(r+\frac{1}{2}\sigma^2)(T-t)+\gamma\sigma\sqrt{T-t}}{\sigma\sqrt{T-t}}$	$d_1 - \sigma\sqrt{T-t}$
b_γ (P)	$e^{-r(T-t)}K\Phi(d_2) - S_t e^{\gamma\sigma\sqrt{T-t}}\Phi(d_1)$	$\frac{\ln\frac{K}{S_t}-(r+\frac{1}{2}\sigma^2)(T-t)-\gamma\sigma\sqrt{T-t}}{\sigma\sqrt{T-t}}$	$d_1 + \sigma\sqrt{T-t}$
a_γ (P)	$e^{-r(T-t)}K\Phi(d_2) - S_t e^{-\gamma\sigma\sqrt{T-t}}\Phi(d_1)$	$\frac{\ln\frac{K}{S_t}-(r+\frac{1}{2}\sigma^2)(T-t)+\gamma\sigma\sqrt{T-t}}{\sigma\sqrt{T-t}}$	$d_1 + \sigma\sqrt{T-t}$

of put option:

$$a_\gamma(P) = e^{r(T-t)}K\left(1 - \Phi\left[\frac{\ln\frac{K}{S_t} - \left(r - \frac{1}{2}\sigma^2\right)(T-t) + \gamma\sigma\sqrt{T-t}}{\sigma\sqrt{T-t}}\right]\right)$$
$$- S_t e^{\gamma\sigma\sqrt{T-t}} \cdot \Phi\left[\frac{\ln\frac{K}{S_t} - (r + \frac{1}{2}\sigma^2)(T-t) + \gamma\sigma\sqrt{T-t}}{\sigma\sqrt{T-t}}\right].$$

If we do the above calculation for the ask price of the European call option and also the bid price of the European put option, we will get the formulas as presented in Table 4.1.

4.4 Data and Numerical Application

In this part, we employ the formulas in Table 4.1 to real data and derive the daily values of the γ. We use the data of European put and call options written on *S&P* 500 index. This index, depicted in Fig. 4.1, mainly consists of large U.S. companies and trade either on the New York Stock Exchange or on the NASDAQ. We chose this index option, because it gives a general overview of the North-American option market that is one of the effective option markets in the world. Additionally, due to the nature of the *S&P* 500 index we may assume that the company specific events will have minor effects on the calculation of values of γ, and all other effects will be the results or aftermath of the financial bubble, when the *S&P* 500 index has declined from its peak around 30% (from 1282 on August 28, 2008, to 900 on October 13, 2008).

We obtained the data from a financial database of Bloomberg Terminal. The data contain the daily bid and ask prices of European call and European put options in the period between 02/01/2008 and 16/12/2010, when U.S. mortgage crises emerged. Both types of options have a strike price of 1300, a base date in January 2008 and an expiration date in December 2010. The data also contain the daily values of the

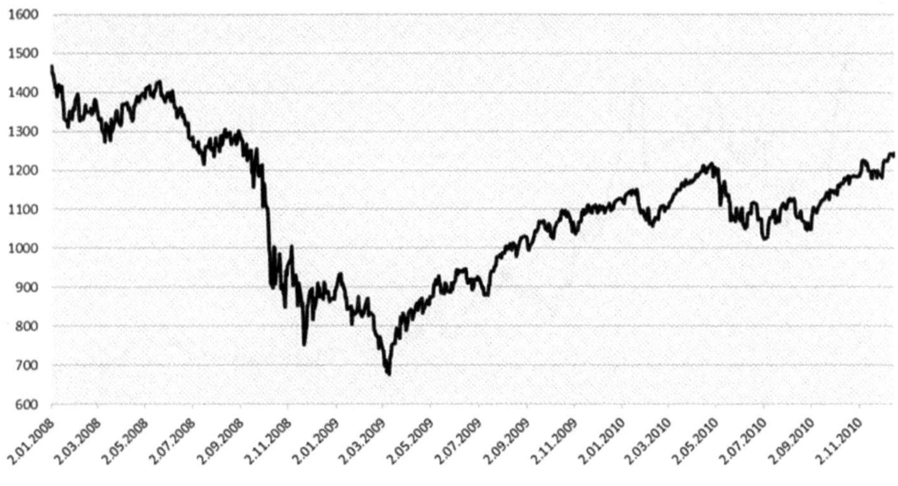

Fig. 4.1 *S&P* 500 index price

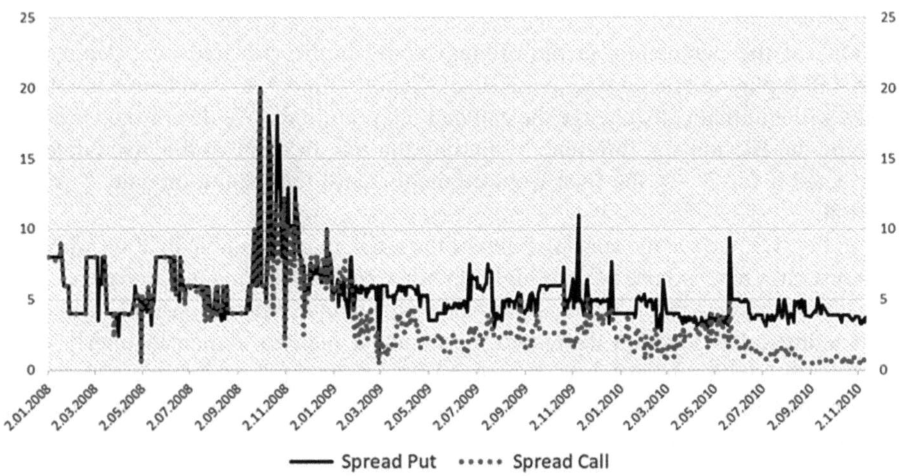

Fig. 4.2 Spreads of the European call and put options

S&P 500 index, quarterly Federal Reserve interest rates, implied volatility values and cover a period of 748 trading days.

As we have mentioned above, there are different distortion functions that can be employed when applying the theory of Conic Finance. In this thesis, we used the Wang transform, because this distortion function is almost similar to *minmaxvar*, which is mostly used in literature, as the corresponding risk measure of *minmaxvar* works best in most cases [2].

In Fig. 4.2, we display the spread between the bid and ask prices of the European Call and Put options. It is obvious that the spread has substantially increased during the U.S. Mortgage bubble.

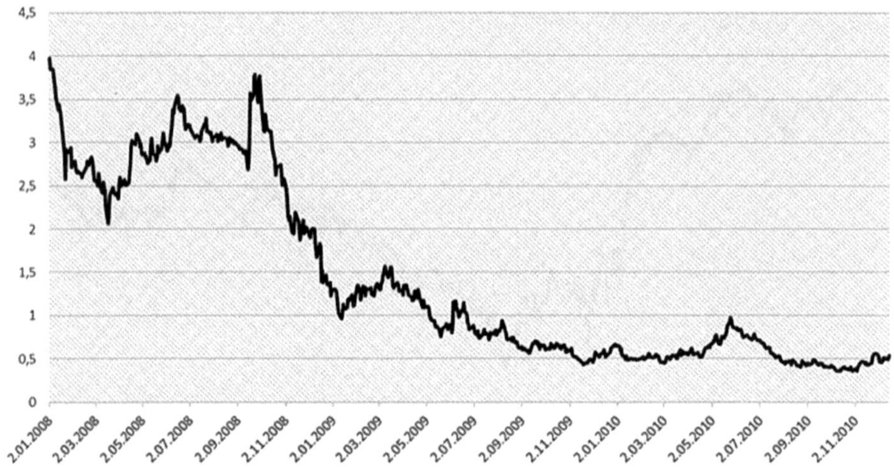

Fig. 4.3 Interest rate on 3 months Treasury Bills

One of the parameters in the pricing model is the risk-free rate, depicted in Fig. 4.3. It plays a crucial role in the Black-Scholes model, as the change in interest rates will automatically change the call and put premiums. We determined r at time t using the Bloomberg Terminal, by finding the risk-free rate at the given date and corrected it for $T - t$, the time from moment t until the expiration date, T, of the option.

In Fig. 4.3, we see the starting point of the stock market bubble in 2008 when the interest rates reached one of their highest points. In 2008, during the mortgage crisis, the labor market weakened and, moreover, business investment, consumer spending, and industrial production decreased, as financial markets remained utterly tense and credit conditions stiffened. In general, the prospect for economic activity has diminished further. Under these circumstances, the Federal Reserve employed all tools and policies to encourage the continuation of sustainable economic growth and to protect price stability. One of the implemented policies was cutting or decreasing the interest rate to the levels that will stimulate the economic activity.

Another parameter, σ, illustrated in Fig. 4.4, is the volatility of the returns on the $S\&P$ 500 index. We make use of implied volatility obtained with the use of Black-Scholes formula.

The values of γ, related to the dataset are estimated by minimizing the total-squared error (TSE). TSE is the sum of the squared variation between the market prices and the theoretical prices:

$$TSE_{bid}(\gamma) = \sum_{i=1}^{\tau} \left(bid_i - b_{\gamma,i}\right)^2, \tag{4.20}$$

$$TSE_{ask}(\gamma) = \sum_{i=1}^{\tau} \left(ask_i - a_{\gamma,i}\right)^2, \tag{4.21}$$

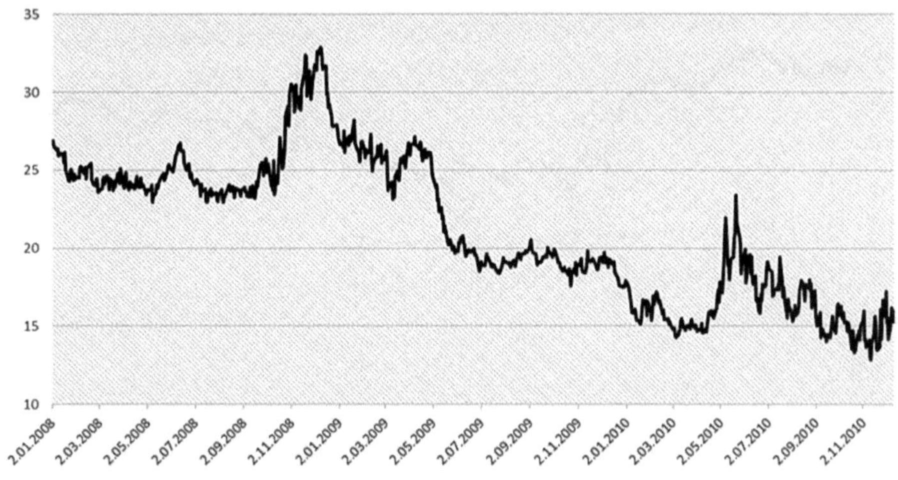

Fig. 4.4 Implied volatility of *S&P* 500 index

or

$$TSE_{bid,ask}(\gamma) = \sum_{i=1}^{\tau} \left(\left(bid_i - b_{\gamma,i}\right)^2 + \left(ask_i - a_{\gamma,i}\right)^2 \right), \tag{4.22}$$

with τ being the number of days for which illiquidity premium will be calculated, for example for daily γ calculations $\tau = 1$, for weekly γ calculation $\tau = 5$, and so on. Minimizing TSE gives the market level's γ. We use (4.22) and the minimization problem can be given as in below:

$$\begin{aligned} \underset{\gamma}{\text{minimize}} \quad & TSE_{bid,ask}(\gamma) \\ \text{subject to} \quad & \gamma \geq 0. \end{aligned} \tag{4.23}$$

The variable γ, can be estimated hourly, daily, weekly, monthly or yearly as a constant value. We have estimated daily γ values. One of the interesting characteristics of the dataset is that it covers the period of the latest financial crises. It is essential to know the effects of a crisis on the acceptability level of the derivative that needs to be priced. These effects can be taken into consideration by adjusting the expected level of the γ.

Analyzing the bid and ask prices of the options and the corresponding values of the γ shows some interesting findings. These values of γ can be used to estimate future values of the γ to price financial products. Figure 4.5 shows the minimum of the TSE of the bid and ask prices of the call option. First of all it is very remarkable that the γ never take values of zero, because this indicates that the bid and ask prices never equal to the Black-Scholes price.

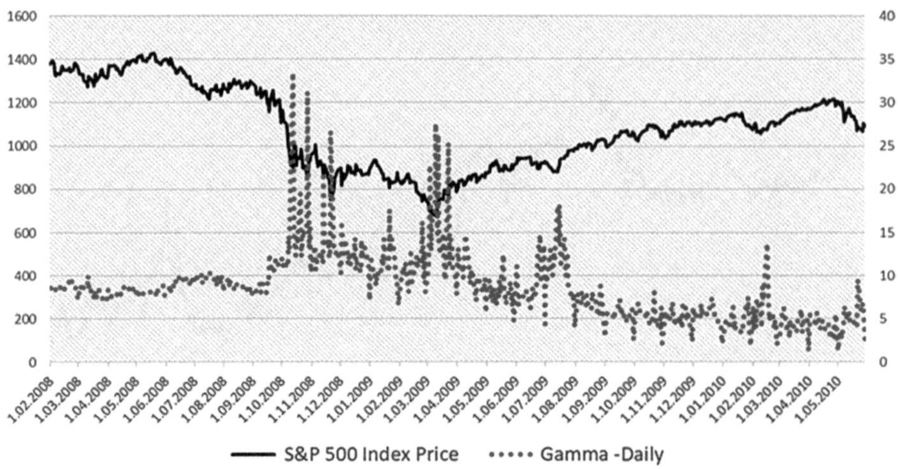

Fig. 4.5 Daily implied illiquidity

The implied illiquidity parameter, or γ, is highest between September 2008 and December 2008. September 2008 was a month full of escalations of several problems in the financial world. Numerous large and important banks had huge problems, because the interbank lending market got severely worse; moreover, the mortgage bubble had already been inflated to its maximum and the burst started. On September 15-th the important investment bank Lehman Brothers went bankrupt after the U.S. government decided not to bail them out, and in the rest of September several banks from all over the world went bankrupt. This led to extreme uncertainty on the financial markets and stock exchanges crashed, which caused higher spreads between bid and ask prices. Therefore, days with a larger bid-ask spread were likely to have higher values for γ.

Other peaks in the γ level in Fig. 4.5 occurred in Spring and Summer 2009 and in Spring 2010. In 2009 there was still uncertainty in the market after the crash in 2008 and important economic powers from all over the world were reporting highly negative economic growth. Furthermore, large companies like Chrysler and General Motors were in financial trouble. This severe decline in economic growth can be an explanation for the peaks of the γ in 2009. The rise of the γ in Spring 2010 can be explained by the European Debt Crisis and by the problems of public finance in countries such as Greece and Spain [6, 7].

References

1. F. Black, M. Scholes, The pricing of options and corporate liabilities. J. Polit. Econ. **81**(3), 637–654 (1973)
2. A. Cherny, D. Madan, New measures for performance evaluation. Rev. Financ. Stud. **22**(7), 2571–2606 (2009)

3. N. Chriss, *Black Scholes and Beyond: Option Pricing Models* (McGraw-Hill, New York, 1996)
4. R. Cont, P. Tankov, *Financial Modelling with Jump Processes* (Chapman & Hall, Boca Raton, 2004)
5. N. Gisiger, Risk-neutral probabilities explained (October 10, 2010). Available at SSRN: https://ssrn.com/abstract=1395390 or http://dx.doi.org/10.2139/ssrn.1395390
6. A.Y. Huang, To what extent does the European debt crisis affect both the European and the global economies, University of Kentucky UKnowledge, 2013
7. R. Kahn, What is the impact of the Eurozone crisis on the U.S. economy? (2013), http://tinyurl.com/jl92yup
8. I. Karatzas, S. Shreve, *Brownian Motion and Stochastic Calculus*, vol. 113 (Springer Science & Business Media, Berlin, 2012)
9. S.I. Resnick, *Adventures in Stochastic Processes* (Springer Science & Business Media, Berlin, 2013)
10. S.E. Shreve, *Stochastic Calculus for Finance II: Continuous-Time Models*, vol. 11 (Springer Science & Business Media, Berlin, 2004)

Chapter 5
Stock Prices Follow a Double Exponential Jump-Diffusion Model

Leptokurticity, volatility clustering as well as implied volatility smile are keywords which represent three deficiencies of Black-Scholes model. Hence, there have been conducted various researches in order to adjust Black-Scholes model and explain these facts on deficiency. Some of those studies include Chaos Theory and fractal Brownian Motion [22, 25], Generalized Hyperbolic models [2, 3, 26], Levy Processes [29], Constant Elasticity Variance model [7, 8] and many more.

In this chapter, we are going to review and apply the Jump-Diffusion models proposed by Kou in [19].

5.1 Details of Jump-Diffusion Models

An experimental interest to use jump-diffusion models are the result of the reality where asset return follow a heavy tail distribution. Nonetheless, heaviness of the tail distributions are ambiguous, while several academicians prefer power type distributions, and some favor exponential-type distributions. It is hard to differentiate power-type tails from exponential-type tails by using empirical data only when one has quite large sample size and a preference of either one of the models as a subjective issue. Moreover, distribution selection has an important meaning in reaction to specifying suitable risk measures [15].

There are numerous methods for the selections of the pricing measure, as jump-diffusion models can result with incomplete markets, some of them are entropy methods, mean-variance hedging, indifference pricing, local mean variance hedging, etc. In this thesis, we employ the Esscher transform, which is giving an easy transition from the base to a risk-neutral probability.

A. Karimov, *Identifying Stock Market Bubbles*, Contributions to Management Science, https://doi.org/10.1007/978-3-319-65009-8_5

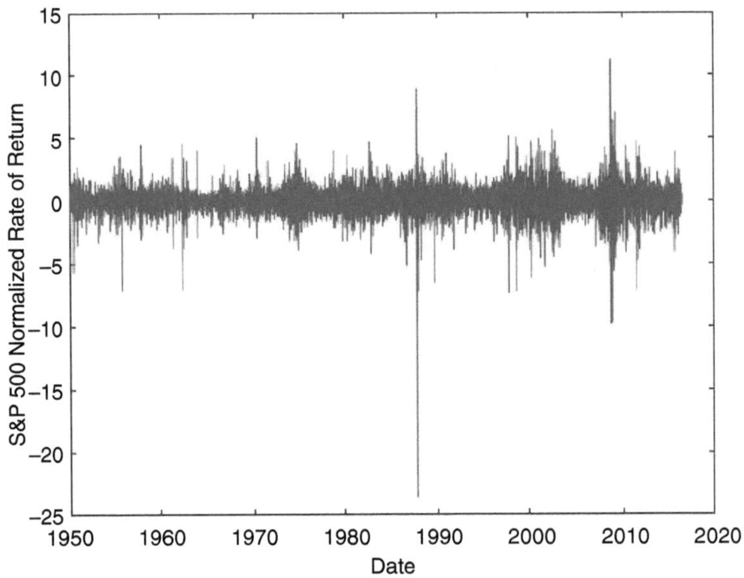

Fig. 5.1 Normalized daily rate of returns of the *S&P* 500 index

5.1.1 Reasons for Using Jump-Diffusion Models

Let us examine the daily prices of *S&P* 500 index (*S&P* 500) from January 1950 until mid 2016. We calculate the daily returns of *S&P* 500 by employing continuously compounded returns, $r_t = \ln \frac{S(t)}{S(t-1)}$ $(t \geq 1)$. The normalized daily rates of returns are plotted in Fig. 5.1.

From Fig. 5.1 we can observe the big spikes in late 80th, 2000 and 2008. The maximum and minimum values are about 11.2422 and −23.5872 times of standard deviation. We have to take into account that for a standard normal random variable Z, $\mathbb{P}(Z < -21.1550) \approx 2.6 \times 10^{-123}$.

Next, let us draw the histogram of the daily returns of *S&P* 500. Figure 5.2 presents the histogram together with the standard normal density function, which is basically constrained inside the interval of $(-3, 3)$. From the figure we may indicate that the features of a high peak and of two heavy tails, of leptokurticity, are quite evident.

5.1.2 Leptokurticity of Returns

Clearly the histogram of *S&P* 500 shows asymmetric heavy tails and a high top. Beside *S&P* 500 this fact is true for almost all asset classes. High peaks and asymmetric heavy tails are so apparent that a name “leptokurtic distribution” is

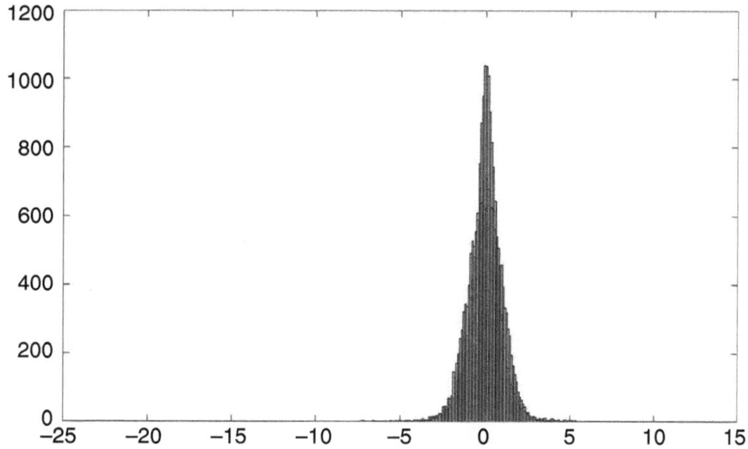

Fig. 5.2 Illustration of the histogram of the normalized daily returns of *S&P* 500 index and the histogram of the random variable from standard normal distribution

used, meaning that there is a large kurtosis. In detail, the kurtosis and skewness are determined as $K = E(\frac{(X-\mu)^4}{\sigma^4})$, and $S = E(\frac{(X-\mu)^3}{\sigma^3})$, respectively, and for the standard normal density the kurtosis is equal to 3. If kurtosis is bigger than 3, the distribution will be named leptokurtic and it will have a higher peak and heavier tails. Double-exponential distribution (*DED*) can be given as an example of leptokurtic distributions.

To evaluate the sample skewness and kurtosis, the following formulas are used:

$$\hat{S} = \frac{1}{(n-1)\hat{\sigma}^3} \sum_{i=1}^{n} (X_i - \bar{X})^3,$$

$$\hat{K} = \frac{1}{(n-1)\hat{\sigma}^4} \sum_{i=1}^{n} (X_i - \bar{X})^4,$$

with $\hat{\sigma}$ being the sample standard deviation. For the daily returns of the *S&P* 500 index, the kurtosis is equal to 30.1178, and the skewness is equal to -1.0101, where negativity of skewness means that the return has a heavier left tail.

5.1.3 *Exponential and Power-Type Tails*

From what we said above, heavy tail distributions of stock returns are very much obvious. In this regard, power and exponential-type tails distributions are the two main classes used in the literature [10]. As indicated by Kou [19], right tails having power-type distributions are not appropriate for the models with continuous compounding.

5.1.4 Implied Volatility Smile

We can easily specify an inverse function that maps option price to the volatility, due to the simple fact that prices derived from Black–Scholes model are monotonically increasing functions of the volatility. In detail, implied volatility $\sigma(T, K)$ is associated with a strike price of option, K, and a maturity T such that if used in the Black–Scholes formula we will achieve a price that accurately equals the market price of an option.

Implied volatility can be easily calculated from real-world prices with diverse maturities and strike prices. If the assumptions of the geometric Brownian motion are true, then implied volatilities should be equal to each other for all the options on the same underlying. Nevertheless, options on the same underlying but with distinctive strikes or maturities will have different implied volatilities.

Particularly, it is extensively investigated that if we depict implied volatilities and strikes, then the volatility curve will resemble a "smile", which means that the implied volatility is a convex function of the strikes. Moreover, the "smile" curve adjusts for distant maturities. It is beneficial to note that leptokurticity under a risk-neutral measure leads to the "volatility smiles" in option prices.

5.1.5 Alternatives for Black-Scholes Model

There have been made numerous analyses in order to customize the Black–Scholes model and include leptokurticity, implied volatility smile and volatility clustering.

One type of these models is fractal Brownian motions, where Brownian motion is replaced by the fractal Brownian motion which has dependent increments [22]. But, Rogers [25] indicates that these types of models may lead to arbitrage opportunities. Another model is given by generalized hyperbolic models, which changes the normal distribution assumption by some other class of Lévy process distributions [2, 3, 26]. There are also stochastic volatility and GARCH models [9, 11, 13, 16], designed to catch the volatility clustering effect. Constant elasticity of variance (CEV) model [7, 8] and time-changed Lévy processes and Brownian motion [4, 5, 14, 20, 21] are other types of alternatives of the aforementioned models.

Yet another type of model that capture leptokurticity, volatility clustering and implied volatility smile is Jump-Diffusion models. Suggested by Merton [23] and Kou [19], jump-diffusion models obey the following representation:

$$S(t) = S(0)e^{(\mu - \frac{1}{2}\sigma^2)t + \sigma W(t)} \prod_{i=1}^{N(t)} e^{Y_i} \quad (t \geq 0),$$

with $(N_t)_{t \geq 0}$ being a Poisson process. The model suggested by Merton assumes that Y follow a normal distribution, and in the model proposed by Kou Y has a DED. This distribution allows to have analytical solutions for some path-dependent options.

5.1.6 *Unique Characteristics of Jump Diffusion*

The asset price process, $(S_t)_{t\geq 0}$, determined under the probability measure $\mathbb{P}$ can be modeled as:

$$\frac{dS_t}{S_{t-}} = \mu dt + \sigma dB_t + d\left(\sum_{i=1}^{N_t}(V_i - 1)\right) \quad (t > 0), \tag{5.1}$$

with $(B_t)_{t\geq 0}$ being a standard Brownian motion, $(N_t)_{t\geq 0}$ Poisson process with rate of λ, and $(V_i)_{i\in\mathbb{N}}$ is a sequence of i.i.d. non-negative random variables. Moreover, all random processes, $(N_t)_{t\geq 0}$, $(B_t)_{t\geq 0}$, and $(V_t)_{t\geq 0}$'s, are pairwise independent. If solved the stochastic differential equation given by Eq. (5.1) will give us the following formula of the asset price process:

$$S_t = S_0 e^{\left(\mu - \frac{1}{2}\sigma^2\right)t + \sigma W(t)} \prod_{i=1}^{N_t} V_i \quad (t \geq 0). \tag{5.2}$$

Kou [19] proposed $Y = \ln(V)$ to have an asymmetric DED with the density

$$f_Y(y) = p\eta_1 e^{-\eta_1 y} 1_{y\geq 0} + q\eta_2 e^{\eta_2 y} 1_{y<0},$$

with $p, q \geq 0$, $p + q = 1$, representing the likelihood of up and down jumps and $\eta_1 > 1$, $\eta_2 > 0$. The prerequisite of $\eta_1 > 1$ is required to guarantee that $E(V) < \infty$ and $E(S(t)) < \infty$, basically meaning that the typical up-jump will not be more than 100%. For clarity and in order to have systematic solutions of option pricing problems, the drift and the volatility terms are considered fixed, and the Brownian motion and jumps are expected to be one-dimensional. Ramezani and Zeng [24] individually suggest a double-exponential jump-diffusion model from an econometric aspect as a way to improve the empirical fit of Merton's normal jump-diffusion model into stock price data.

The DED has two appealing characteristics that are important for the model. First of all, it has the leptokurticity [17]. The leptokurticity of the distribution of jump sizes is acquired by the return distribution. Secondly, as a special property of the DED, originated from the exponential distribution, is the memoryless property. This unique feature clarifies why closed-form solutions for option pricing problems under the double-exponential jump-diffusion are possible even though it sounds complicated for other models.

5.2 Jump-Diffusion Model

Assume that we have the following process given by the Stochastic Differential Equation (SDE),

$$\frac{dS(t)}{S(t^-)} = \mu dt + \sigma dW(t) + d\sum_{i=1}^{N(t)}(V_i - 1) \quad (t > 0). \tag{5.3}$$

We will use the Lévy-Itô Theorem in order to solve the SPDE.

Theorem 5.1 (Lévy-Itô [28]) *For a process of the type*

$$L_t = L_0 + \int_0^t b_s ds + \int_0^t \sigma_s dW_s + \sum_{i=1}^{N_t} \Delta X_i, \tag{5.4}$$

where $t \geq 0$, $(W_s)_{s\geq 0}$ *is a Standard Brownian Motion, and* $(N_t)_{s\geq 0}$ *is the Poisson Process, the Itô Formula will have the following representation:*

$$df(L_t, t) = \frac{\partial f(L_t, t)}{\partial t}dt + b_t\frac{\partial f(L_t, t)}{\partial x}dt + \frac{\sigma_t^2}{2}\frac{\partial^2 f(L_t, t)}{\partial x^2}dt + \sigma_t\frac{\partial f(L_t, t)}{\partial x}dW_t + [f(X_{t^-} + \Delta X_t) - f(X_{t^-})] . \tag{5.5}$$

By employing the Itô Formula from Eq. (5.5) to the log returns of a price process, $f(S_t) = \ln(S_t)$, we will gradually obtain the subsequent expressions:

$$df = \frac{\partial f}{\partial t}dt + \mu S_t\frac{\partial f}{\partial S_t}dt + \frac{\sigma^2 S_t^2}{2}\frac{\partial^2 f}{\partial S_t^2}dt + \sigma S_t\frac{\partial f}{\partial S_t}dW_t + [f(S_{t^-} + S_t(V_t - 1)) - f(S_{t^-})] ,$$

by making a substitute of $f(S_t) = \ln S_t$, we will have:

$$d(\ln S_t) = \mu S_t\frac{1}{S_t}dt - \frac{\sigma^2 S_t^2}{2}\frac{1}{S_t^2}dt + \sigma S_t\frac{1}{S_t}dW_t + \ln S_t + \ln V_t - \ln S_t,$$

or

$$d(\ln S_t) = (\mu - \frac{\sigma^2}{2})dt + \sigma dW_t + \ln V_t.$$

By integrating both sides of the above Stochastic Differential Equation we will have the following equation,

$$\ln S_t = \ln S_0 + (\mu - \frac{\sigma^2}{2})t + \sigma W_t + \sum_{k=1}^{N(t)} \ln V_t,$$

or

$$S_t = S_0 e^{(\mu - \frac{\sigma^2}{2})t + \sigma W_t + \sum_{k=1}^{N(t)} Y_k},$$

where $Y_k = \ln V_t$.

Finally, we will have an exponential Lévy process $S_t = S_0 e^{L_t}$ $(t \geq 0)$ where L_t has the following representation:

$$L_t = \underbrace{\left(\mu - \frac{\sigma^2}{2}\right) t}_{\text{Drift}} + \underbrace{\sigma W_t}_{\text{Brownian Motion}} + \underbrace{\sum_{k=1}^{N(t)} Y_k}_{\text{Sum of Double-Exponential process}} \qquad (t \geq 0). \tag{5.6}$$

The distribution of the sum of Drift and Brownian Motion parts has a Normal Distribution with the mean of $\left(\mu - \frac{\sigma^2}{2}\right) t + \ln S_0$, where S_0 is the initial value of the price process, and with the standard deviation of $\sigma\sqrt{t}$.

Now let us calculate the distribution function of Sum of Double-Exponential process and use the Convolution Formula [18] to combine the above two distributions in order to find the distribution of $(L_t)_{t \geq 0}$.

5.3 Distribution Function of Jump Process

In order to calculate the distribution of the sum of the Double-Exponential process, we will employ the characteristics function. From probability theory we know that the characteristics function φ of the distribution f is given as,

$$\varphi_X(u) = \int_{-\infty}^{\infty} e^{iux} f(x) dx. \tag{5.7}$$

We have the following distribution:

$$f_X(x) = p\eta_1 e^{-\eta_1 x} 1_{x \geq 0} + q\eta_2 e^{\eta_2 x} 1_{x<0}, \tag{5.8}$$

where $\eta_1 > 1$ and $\eta_2 > 0$. If we apply Eq. (5.7) to the distribution function in Eq. (5.8), then we will have the following representation:

$$\begin{aligned}\varphi_X(u) &= \int_{-\infty}^{\infty} e^{\mathrm{i}ux} f(x)dx = \int_{-\infty}^{\infty} e^{\mathrm{i}ux}\left[p\eta_1 e^{-\eta_1 x} 1_{x\geq 0} + q\eta_2 e^{\eta_2 x} 1_{x<0}\right] dx \\ &= p\eta_1 \int_0^{\infty} e^{-x(\eta_1-\mathrm{i}u)} dx + q\eta_2 \int_{-\infty}^{0} e^{x(\eta_2+\mathrm{i}u)} dx \\ &= p\eta_1 \left(\frac{e^{-x(\eta_1-\mathrm{i}u)}}{-(\eta_1-\mathrm{i}u)}\Big|_0^{\infty}\right) + q\eta_2 \left(\frac{e^{x(\eta_2+\mathrm{i}u)}}{\eta_2+\mathrm{i}u}\Big|_{-\infty}^{0}\right) \\ &= \frac{p\eta_1}{\eta_1-\mathrm{i}u} + \frac{q\eta_2}{\eta_2+\mathrm{i}u}.\end{aligned}$$

Let us start our calculations with the below transformation,

$$\begin{aligned}\varphi_{\sum_{k=1}^{N_t} X_k}(u) &= E\left\{e^{\mathrm{i}u\sum_{k=1}^{N_t} X_k}\right\} \\ &= \sum_{n=1}^{\infty} E\left\{e^{\mathrm{i}u\sum_{k=1}^{N_t} X_k} | N_t = n\right\} \mathbb{P}(N_t = n) \\ &= \sum_{n=1}^{\infty} E\left\{e^{\mathrm{i}u\sum_{k=1}^{n} X_k}\right\} \mathbb{P}(N_t = n) \\ &= \sum_{n=1}^{\infty} \varphi_{\sum_{k=1}^{n} X_k}(u)\, \mathbb{P}(N_t = n).\end{aligned} \tag{5.9}$$

In order to compute the distribution function g of the $\Theta = \sum_{k=1}^{N_t} X_k$ we have to employ the Inversion Formula on the above calculated characteristics function.

Theorem 5.2 (Inversion Formula for Densities [18]) *Let X be a real random variable whose characteristics function φ is integrable over $\mathbb{R}$, so $\int_{-\infty}^{\infty} |\varphi(u)|\, du < \infty$. Then X has a bounded continuous density f on $\mathbb{R}$ given by,*

$$f(x) = \frac{1}{2\pi} \int_{-\infty}^{\infty} e^{-\mathrm{i}ux} \varphi(u) du. \tag{5.10}$$

If we apply Theorem 5.2 to the final outcome of Eq. (5.9), we will get the following expression for the distribution function of Θ:

$$g_\Theta(x) = \frac{1}{2\pi} \int_{-\infty}^{\infty} e^{-\mathrm{i}ux} \sum_{n=1}^{\infty} \left\{\varphi_{\sum_{k=1}^{n} X_k}(u)\, \mathbb{P}(N_t = n)\right\} du. \tag{5.11}$$

Using the Tonelli's theorem [30] on interchanging the sequence of integration and summation and the fact that function of $\varphi_{\sum_{k=1}^{n} X_k}(u)\,\mathbb{P}(N_t = n)$ is non-negative we will be able to make the following changes in Eq. (5.11):

$$g_\Theta(x) = \sum_{n=1}^{\infty} \left\{ \frac{1}{2\pi} \int_{-\infty}^{\infty} e^{-\mathrm{i}ux} \varphi_{\sum_{k=1}^{n} X_k}(u)\,\mathbb{P}(N_t = n) du \right\} \tag{5.12}$$

and as the $\mathbb{P}(N_t = n)$ does not depend from the variable u it can be taken out of the integral. Hence, we will have the following equality,

$$\begin{aligned} g_\Theta(x) &= \sum_{n=1}^{\infty} \left\{ \frac{1}{2\pi} \int_{-\infty}^{\infty} e^{-\mathrm{i}ux} \varphi_{\sum_{k=1}^{n} X_k}(u) du \right\} \mathbb{P}(N_t = n) \\ &= \sum_{n=1}^{\infty} g_{\Theta_n}(x)\mathbb{P}(N_t = n), \end{aligned} \tag{5.13}$$

where $\Theta_n = \sum_{k=1}^{n} X_k$.

Using the final part of Eq. (5.13) we may deduce that, in order to calculate the distribution function $\sum_{k=1}^{N_t} X_k$ we will have to calculate the characteristics function of $\sum_{k=1}^{n} X_k$. Moreover, by using this characteristics function we will derive the distribution function of $\sum_{k=1}^{n} X_k$ and afterward we will use Eq. (5.13) to arrive at the distribution function of $\sum_{k=1}^{N_t} X_k$. In this regards, we have the condition that all the X_k, $k \in \mathbb{N}$, are i.i.d. random variables. Hence, the characteristics functions of $\Theta_n = \sum_{k=1}^{n} X_k$, being sums of aforementioned random variables, would equate to the product of the characteristic functions, or:

$$\begin{aligned} \varphi_{\sum_{k=1}^{n} X_k}(u) = \prod_{k=1}^{n} \varphi_{X_k}(u) &= \prod_{k=1}^{n} \left[\frac{p\eta_1}{\eta_1 - \mathrm{i}u} + \frac{q\eta_2}{\eta_2 + \mathrm{i}u} \right] \\ &= \left(\frac{p\eta_1}{\eta_1 - \mathrm{i}u} + \frac{q\eta_2}{\eta_2 + \mathrm{i}u} \right)^n. \end{aligned} \tag{5.14}$$

If we apply Theorem 5.2 to Eq. (5.14), we will get the following expression for the distribution function of $(\Theta_n)_{n\in\mathbb{N}}$:

$$g_{\Theta_n}(x) = \frac{1}{2\pi} \int_{-\infty}^{\infty} e^{-\mathrm{i}ux} \left(\frac{p\eta_1}{\eta_1 - \mathrm{i}u} + \frac{q\eta_2}{\eta_2 + \mathrm{i}u} \right)^n du.$$

For the cases of $n \geq 2$,

$$g_{\Theta_n}(x) = \frac{1}{2\pi} \lim_{R \to +\infty} \int_{-R}^{R} e^{-\mathrm{i}ux} \left(\frac{p\eta_1}{\eta_1 - \mathrm{i}u} + \frac{q\eta_2}{\eta_2 + \mathrm{i}u} \right)^n du$$
$$= \frac{1}{2\pi} \lim_{R \to +\infty} \sum_{k=0}^{n} \binom{n}{k} (p\eta_1)^k (q\eta_2)^{n-k} \int_{-R}^{R} \frac{e^{-\mathrm{i}ux}}{(\eta_1 - \mathrm{i}u)^k (\eta_2 + \mathrm{i}u)^{n-k}} du.$$

If we change the places of summation and limit, then we will get the following equation:

$$g_{\Theta_n}(x) = \frac{1}{2\pi} \sum_{k=0}^{n} \binom{n}{k} (p\eta_1)^k (q\eta_2)^{n-k} \lim_{R \to +\infty} \int_{-R}^{R} \frac{e^{-\mathrm{i}zx}}{(\eta_1 - \mathrm{i}z)^k (\eta_2 + \mathrm{i}z)^{n-k}} dz. \tag{5.15}$$

Let us calculate the function $g_{\Theta_n}(x)$ for the case $x \geq 0$.
We introduce a function G_k by

$$G_k(z, x) = \frac{e^{-\mathrm{i}zx}}{(\eta_1 - \mathrm{i}z)^k (\eta_2 + \mathrm{i}z)^{n-k}}, \tag{5.16}$$

and let $D_R = \breve{ABCD}$ be a curve connecting points of A, B, C, D depicted in Fig. 5.3.

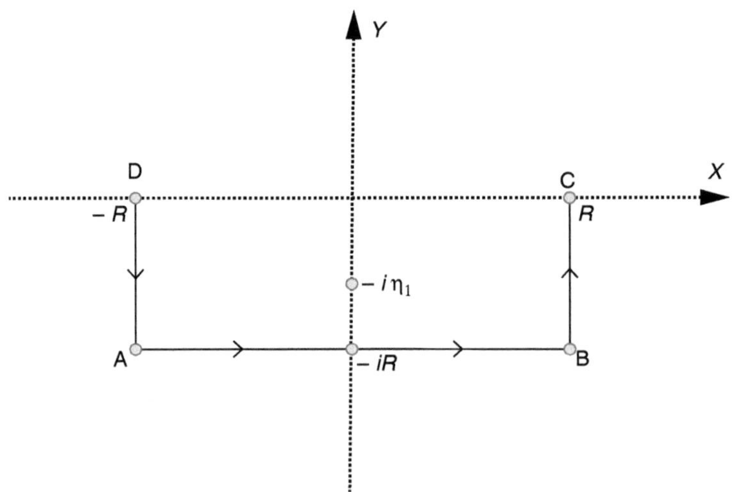

Fig. 5.3 The point of singularity and the curve of integration of the function G_k

It can easily be seen that for large enough values of $R > 0$ in Fig. 5.3, the following equality holds:

$$\begin{aligned}-\int_{-R}^{R} G_k(z,x)dz = \int_{D_R} G_k(z,x)dz - \int_{\overset{\smile}{DA}} G_k(z,x)dz \\ - \int_{\overset{\smile}{AB}} G_k(z,x)dz - \int_{\overset{\smile}{BC}} G_k(z,x)dz.\end{aligned} \tag{5.17}$$

The equation of the curve of $\overset{\smile}{DA}$ is $z = -R - i\eta$ for all $0 \le \eta \le R$. Hence, for each $z \in \overset{\smile}{DA}$ and for sufficiently large values of R, we will get the following expression:

$$\begin{aligned}\left|(\eta_1 - \mathrm{i}z)^k(\eta_2 + \mathrm{i}z)^{n-k}\right| &\ge \left||z| - \eta_1\right|^k \left||z| - \eta_2\right|^{n-k} \\ &\ge (R - \eta_1)^k \, (R - \eta_2)^{n-k} \\ &\ge \left(\frac{R}{2}\right)^k \left(\frac{R}{2}\right)^{n-k} = \left(\frac{R}{2}\right)^n.\end{aligned} \tag{5.18}$$

The equation of the curve of $\overset{\smile}{AB}$ is $z = \xi - iR$ for all $-R \le \xi \le R$. Hence, for all $z \in \overset{\smile}{AB}$ and for sufficiently big values of R, the relationship of (5.18) holds.

The equation of the curve of $\overset{\smile}{BC}$ is $z = R + i\eta$ for all $-R \le \eta \le 0$. Hence, for all $z \in \overset{\smile}{BC}$ and for sufficiently large values of R, the relationship of (5.18) holds.

Moreover, as for all $z \in \overset{\smile}{ABCD}$ we have $\operatorname{Im} Z \le 0$, then for all $z \in \overset{\smile}{ABCD}$ and $x \ge 0$ the following inequality holds:

$$\left|e^{-\mathrm{i}zx}\right| = e^{x \operatorname{Im} z} \le 1. \tag{5.19}$$

From (5.16), (5.18), and (5.19), for all $z \in \overset{\smile}{DA} \cup \overset{\smile}{AB} \cup \overset{\smile}{BC}$ and $x \ge 0$, the following estimation will be satisfied:

$$|G_k(z,x)| \le \left(\frac{2}{R}\right)^n, \tag{5.20}$$

where $k = 0, 1, \ldots, n$ $(n \ge 2)$ and R is sufficiently large number.

In accordance with the inequality (5.20), the following calculations hold true:

$$\begin{aligned}
\left|\int_{\breve{DA}} G_k(z,x)dz\right| &\le \left(\frac{2}{R}\right)^n \cdot \left|\breve{DA}\right| = \frac{2^n}{R^n}\cdot R = \frac{2^n}{R^{n-1}},\\
\left|\int_{\breve{AB}} G_k(z,x)dz\right| &\le \left(\frac{2}{R}\right)^n \cdot \left|\breve{AB}\right| = \frac{2^n}{R^n}\cdot 2R = \frac{2^{n+1}}{R^{n-1}},\\
\left|\int_{\breve{BC}} G_k(z,x)dz\right| &\le \left(\frac{2}{R}\right)^n \cdot \left|\breve{BC}\right| = \frac{2^n}{R^n}\cdot R = \frac{2^n}{R^{n-1}},
\end{aligned} \tag{5.21}$$

where $|l|$ indicates the length of the curve of l.

From here we can easily get the following equalities,

$$\lim_{R\to+\infty}\int_{\breve{DA}} G_k(z,x)dz = \lim_{R\to+\infty}\int_{\breve{AB}} G_k(z,x)dz = \lim_{R\to+\infty}\int_{\breve{BC}} G_k(z,x)dz = 0. \tag{5.22}$$

Taking into account (5.22) in (5.17), we will find that,

$$\lim_{R\to+\infty}\int_{-R}^{R} G_k(z,x)dz = -\lim_{R\to+\infty}\int_{D_R} G_k(z,x)dz. \tag{5.23}$$

From Eqs. (5.23), (5.15) and (5.16) we will get the subsequent expression for $g_{\Theta_n}(y)$:

$$g_{\Theta_n}(x) = -\frac{1}{2\pi}\sum_{k=0}^{n}\binom{n}{k}(p\eta_1)^k(q\eta_2)^{n-k}\lim_{R\to+\infty}\int_{D_R} G_k(z,x)dz. \tag{5.24}$$

To calculate the integral at the right-hand side of (5.24), we will benefit from the theorems of Residue Theory [27].

Since, the function $G_k(z,x)$ is analytical with respect to the variable z in the domain, which is bounded by curve of D_R; hence, the following equality holds

$$\int_{D_R} G_0(z,x)dz = 0. \tag{5.25}$$

Thus, we may infer that,

$$g_{\Theta_n}(x) = -\frac{1}{2\pi}\sum_{k=1}^{n}\binom{n}{k}(p\eta_1)^k(p\eta_2)^{n-k}\lim_{R\to+\infty}\int_{D_R} G_k(z,x)dz. \tag{5.26}$$

For $k = 1, 2, \ldots, n$, the function $G_k(z,x)$ is meromorphic with respect to the z in the domain, bounded with curve D_R, and the only singularity point is $z = -i\eta_1$.

Therefore, in accordance with the Residue Theory we will have the following expression for G_k for $k = 1, 2, \dots, n$,

$$\int_{D_R} G_k(z,x)dz = 2\pi \mathrm{i} \cdot \underset{z=-\mathrm{i}\eta_1}{Res} G_k(z,x).$$

By this way (5.26) will be simplified to the following term,

$$g_{\Theta_n}(x) = -\mathrm{i}\sum_{k=1}^{n} \binom{n}{k} (p\eta_1)^k (p\eta_2)^{n-k} \lim_{R\to+\infty} \underset{z=-\mathrm{i}\eta_1}{Res} G_k(z,x). \tag{5.27}$$

It is known that [27], if the function $f(z)$ is analytical around the point a, then

$$\underset{z=a}{Res} \frac{f(z)}{(z-a)^k} = \frac{1}{(k-1)!} \frac{d^{k-1}}{dz^{k-1}} f(z)\bigg|_{z=a}. \tag{5.28}$$

If we write the function $G_k(z,x)$ as

$$G_k(z,x) = \frac{\frac{e^{-\mathrm{i}zx}}{(-\mathrm{i})^k(\eta_2+\mathrm{i}z)^{n-k}}}{(z+\mathrm{i}\eta_1)^k},$$

and use Eq. (5.28), then we will get the subsequent equation:

$$\underset{z=-\mathrm{i}\eta_1}{Res} G_k(z,x) = \frac{\mathrm{i}^k}{(k-1)!} \frac{d^{k-1}}{dz^{k-1}} \left(\frac{e^{-\mathrm{i}zx}}{(\eta_2+\mathrm{i}z)^{n-k}} \right) \bigg|_{z=-\mathrm{i}\eta_1}. \tag{5.29}$$

By this way and in accordance with Eqs. (5.27) and (5.29) we will have,

$$g_{\Theta_n}(x) = -\sum_{k=1}^{n} \frac{\mathrm{i}^{k+1}\binom{n}{k}(p\eta_1)^k(q\eta_2)^{n-k}}{(k-1)!} T_k(x), \tag{5.30}$$

where

$$T_k(x) = \frac{d^{k-1}}{dz^{k-1}} \left(\frac{e^{-\mathrm{i}zx}}{(\eta_2+\mathrm{i}z)^{n-k}} \right) \bigg|_{z=-\mathrm{i}\eta_1} \quad (k = 1, 2, \dots, n). \tag{5.31}$$

For $k = n$, Eq. (5.31) will have the subsequent expression:

$$T_n(x) = \frac{d^{n-1}}{dz^{n-1}} e^{-\mathrm{i}zx} \bigg|_{z=-\mathrm{i}\eta_1} = (-\mathrm{i}x)^{n-1} e^{-\eta_1 x}. \tag{5.32}$$

Moreover, for all $\alpha \in \mathbb{N}$ we have the following representation:

$$\frac{d^j}{dz^j} \left(\frac{1}{(\eta_2+\mathrm{i}z)^{\alpha}} \right) \bigg|_{z=-\mathrm{i}\eta_1} = \frac{(-\mathrm{i})^j(\alpha+j-1)!}{(\alpha-1)!(\eta_1+\eta_2)^{\alpha+j}}. \tag{5.33}$$

Taking into account Eqs. (5.31) and (5.33) for $k = 1, 2, \ldots, n-1$, we will have the subsequent representation for $T_k(x)$:

$$T_k(x) = \sum_{j=0}^{k-1} \binom{k-1}{j} (e^{-\mathrm{i}zx})^{(j)} \left(\frac{1}{(\eta_2 + \mathrm{i}z)^{n-k}} \right)^{(k-1-j)} \Bigg|_{z=-\mathrm{i}\eta_1}$$
$$= \sum_{j=0}^{k-1} \binom{k-1}{j} (-\mathrm{i}x)^j e^{-\eta_1 x} \frac{(-\mathrm{i})^{k-j-1}(n-j-2)!}{(n-k-1)!(\eta_1+\eta_2)^{n-j-1}}.$$

Combining the above result and Eqs. (5.30) and (5.32), we will get the following expressions for $g_{\Theta_n}(x)$. For the case of $x \geq 0$,

$$g_{\Theta_n}(x) = e^{-\eta_1 x} \sum_{k=1}^{n-1} \frac{\mathrm{i}^{k+1}\binom{n}{k}(p\eta_1)^k (q\eta_2)^{n-k}}{(k-1)!} \sum_{j=0}^{k-1} \frac{\binom{k-1}{j}(-\mathrm{i}x)^j(-\mathrm{i})^{k-j-1}(n-j-2)!}{(n-k-1)!(\eta_1+\eta_2)^{n-j-1}}$$
$$- \frac{\mathrm{i}^{n+1}\binom{n}{n}(p\eta_1)^n}{(n-1)!}(-\mathrm{i}x)^{n-1} e^{-\eta_1 x},$$

or,

$$g_{\Theta_n}(x) = \left[\frac{(p\eta_1)^n}{(n-1)!} x^{n-1} + \sum_{k=1}^{n-1} \sum_{j=0}^{k-1} \frac{\binom{n}{k}\binom{k-1}{j}(n-j-2)!(p\eta_1)^k(q\eta_2)^{n-k}}{(k-1)!(n-k-1)!(\eta_1+\eta_2)^{n-j-1}} x^j \right] e^{-\eta_1 x}.$$

The last formula can be rewritten by the subsequent expression:

$$g_{\Theta_n}(x) := g^1_{\Theta_n}(x) = \left[\frac{(p\eta_1)^n}{(n-1)!} x^{n-1} \right.$$
$$\left. + \sum_{j=0}^{n-2} \frac{(n-j-2)!}{(\eta_1+\eta_2)^{n-j-1}} \left(\sum_{k=j+1}^{n-1} \frac{\binom{n}{k}\binom{k-1}{j}(p\eta_1)^k(q\eta_2)^{n-k}}{(k-1)!(n-k-1)!} \right) x^j \right] e^{-\eta_1 x}, \tag{5.34}$$

where $n \geq 2$. We have to note that Eq. (5.34) is correct only for the case of $x \geq 0$.

Let us now calculate the $g_{\Theta_n}(x)$ for the case of $x < 0$, and $n \geq 2$:

$$g_{\Theta_n}(x) = \frac{1}{2\pi} \int_{-\infty}^{+\infty} e^{-\mathrm{i}ux} \left(\frac{p\eta_1}{\eta_1 - \mathrm{i}u} + \frac{q\eta_2}{\eta_2 + \mathrm{i}u} \right)^n du$$
$$\overset{u=-u}{=} \frac{1}{2\pi} \int_{-\infty}^{+\infty} e^{\mathrm{i}ux} \left(\frac{q\eta_2}{\eta_2 - \mathrm{i}u} + \frac{p\eta_1}{\eta_1 + \mathrm{i}u} \right)^n du,$$

or,

$$g_{\Theta_n}(x) = \frac{1}{2\pi}\int_{-\infty}^{+\infty} e^{-iu(-x)}\left(\frac{q\eta_2}{\eta_2 - iu} + \frac{p\eta_1}{\eta_1 + iu}\right)^n du. \tag{5.35}$$

In Eq. (5.35), we refer to $-x > 0$, so we can infer the $g_{\Theta_n}(x)$ formula for the case of $x < 0$ from Eq. (5.34), simply by changing the variables x and η_1 in Eq. (5.34) by $-x$ and η_2, respectively. In this way, we will get the following equation for $g_{\Theta_n}(x)$, if $x < 0$, and $n \geq 2$,

$$g_{\Theta_n}(x) := g^2_{\Theta_n}(x) = \Bigg[\frac{(q\eta_2)^n}{(n-1)!}(-x)^{n-1} + \sum_{j=0}^{n-2}\frac{(n-j-2)!}{(\eta_1+\eta_2)^{n-j-1}}\left(\sum_{k=j+1}^{n-1}\frac{\binom{n}{k}\binom{k-1}{j}(q\eta_2)^k(p\eta_1)^{n-k}}{(k-1)!(n-k-1)!}\right)(-x)^j\Bigg] e^{\eta_2 x}. \tag{5.36}$$

We may combine the two representations of the $g_{\Theta_n}(x)$ distribution function, given by (5.34) and (5.36) into one as follows:

$$g_{\Theta_n}(x) = g^1_{\Theta_n}(x)1_{x\geq 0} + g^2_{\Theta_n}(x)1_{x<0}, \tag{5.37}$$

where

$$g^1_{\Theta_1}(x) = p\eta_1 e^{-\eta_1 x}, g^2_{\Theta_1}(x) = q\eta_2 e^{\eta_2 x}. \tag{5.38}$$

By calculating the functions $g_{\Theta_n}(x)$ we have found the distribution function of sum of n i.i.d. Double Exponential Processes, $\Theta_n = \sum_{k=1}^n X_k$, where $n \in \mathbb{N}$. Let us calculate the distribution function of $\Theta = \sum_{k=1}^{N_t} X_k$. It is well known that $(N_t)_{t\geq 0}$ is a Poisson Process with the following distribution:

$$\mathbb{P}(N_t = n) = e^{-\lambda t}\frac{(\lambda t)^n}{n!} \quad (t \geq 0), \tag{5.39}$$

where any positive real λ is equal to the expected value of N_t and with its variance,

$$t\lambda = E(N_t) = Var(N_t).$$

If we will use the PDF of Poisson distribution given by Eq. (5.39) with Eq. (5.37), we will get the following formula for the distribution function of Θ:

$$g_\Theta(x) = \sum_{n=1}^{\infty}\left[g^1_{\Theta_n}(x)1_{x\geq 0} + g^2_{\Theta_n}(x)1_{x<0}\right]e^{-\lambda t}\frac{(\lambda t)^n}{n!}. \tag{5.40}$$

In order to get the distribution function of L_t, we will use the following definition and theorem [18].

Definition 5.1 ([18]) Let X and Y be two continuous random variables with density functions $f(x)$ and $g(y)$, respectively. We assume that both $f(x)$ and $g(y)$ are defined for all real numbers. Then the *convolution* of f and g, $f * g$, is the function given by

$$\begin{aligned}(f * g)(z) &= \int_{-\infty}^{+\infty} f(z-y)g(y)dy \\ &= \int_{-\infty}^{+\infty} g(z-y)f(y)dy.\end{aligned} \tag{5.41}$$

Theorem 5.3 ([18]) *Let X and Y be two independent random variables with density functions $f_X(x)$ and $f_Y(y)$ defined for all x. Then, the sum $Z = X + Y$ is a random variable with density function $f_Z(z)$, where f_Z is the convolution of f_X and f_Y.*

Let us take a closer look at Eq. (5.6). We may see that, as indicated above, the right side can be separated into two components, one for the Sum of Double-Exponential Process, $g_{\Theta_n}(x)$, and one for the Drift plus Brownian Motion part, $g_{DBM}(x)$. The distribution of the Sum of Double-Exponential Processes is characterized with Eq. (5.37), and the distribution of the Drift plus Brownian Motion part is given subsequently:

$$g_{DBM}(x) = \frac{1}{\sigma\sqrt{2\pi t}} e^{\frac{-\left(x-\left(\left(\mu-\frac{\sigma^2}{2}\right)t+\ln S_0\right)\right)^2}{2\sigma^2 t}}. \tag{5.42}$$

In order to find the distribution of the jump process $(L_t)_{t\geq 0}$ we will use Theorem 5.3, so the distribution would be as follows:

$$\begin{aligned}g_{L_t}(z) &= \int_{-\infty}^{+\infty} g_{DBM}(z-x)g_{\Theta}(x)dx \\ &= \int_{-\infty}^{+\infty} g_{DBM}(z-x)\sum_{n=1}^{\infty}\left[g^1_{\Theta_n}(x)1_{x\geq 0} + g^2_{\Theta_n}(x)1_{x<0}\right]e^{-\lambda t}\frac{(\lambda t)^n}{n!}dx \\ &= \sum_{n=1}^{\infty}\left[\int_0^{+\infty} g_{DBM}(z-x)g^1_{\Theta_n}(x)dx\right]\frac{(\lambda t)^n}{n!}e^{-\lambda t} \\ &\quad + \sum_{n=1}^{\infty}\left[\int_{-\infty}^{0} g_{DBM}(z-x)g^2_{\Theta_n}(x)dx\right]\frac{(\lambda t)^n}{n!}e^{-\lambda t} \\ &= g^1_{L_t}(z) + g^2_{L_t}(z).\end{aligned} \tag{5.43}$$

As already have been mentioned we need to calculate the distribution F_{S_t} of the Stock Price Process given by $S_t = e^{L(t)}$ $(t \geq 0)$.

In order to calculate these distributions, by using Eq. (5.43), we will employ the following transformation:

$$F_{S_t}(x) = \mathbb{P}(S_t \leq x) = \mathbb{P}\left(L_t(z) \leq \ln(\frac{x}{S_0})\right) = \int_{-\infty}^{\ln(\frac{x}{S_0})} g_{L_t}(z)dz = \\ = \int_{-\infty}^{\ln(\frac{x}{S_0})} g^1_{L_t}(z)dz + \int_{-\infty}^{\ln(\frac{x}{S_0})} g^2_{L_z}(z)dz, \tag{5.44}$$

where

$$g^1_{L_t}(z) = \sum_{n=1}^{+\infty}\left[\int_0^{+\infty} g_{DBM}(z-\xi)g^1_{\Theta_n}(\xi)d\xi\right]\frac{(\lambda t)^n}{n!}e^{-\lambda t}, \tag{5.45}$$

and

$$g^2_{L_z}(z) = \sum_{n=1}^{+\infty}\left[\int_{-\infty}^{0} g_{DBM}(z-\xi)g^2_{\Theta_n}(\xi)d\xi\right]\frac{(\lambda t)^n}{n!}e^{-\lambda t}. \tag{5.46}$$

From Eqs. (5.44)–(5.46) we will get,

$$F_{S_t}(x) = \sum_{n=1}^{\infty}\left[\sum_{s=1}^{2} A_n^{(s)}(x)\right]\frac{(\lambda t)^n}{n!}e^{-\lambda t}, \tag{5.47}$$

where

$$A_n^{(1)}(x) = \int_{-\infty}^{\ln(\frac{x}{S_0})} dz \int_0^{+\infty} g_{DBM}(z-\xi)g^1_{\Theta_n}(\xi)d\xi, \tag{5.48}$$

$$A_n^{(2)}(x) = \int_{-\infty}^{\ln(\frac{x}{S_0})} dz \int_{-\infty}^{0} g_{DBM}(z-\xi)g^2_{\Theta_n}(\xi)d\xi. \tag{5.49}$$

We shall calculate the expressions for $A_n^{(1)}(x)$ and $A_n^{(2)}(x)$ separately. Let us make the following designation:

$$\tau = \ln(\frac{x}{S_0}). \tag{5.50}$$

By making a variable change of $\xi = z - \xi$ and using Eq. (5.50) in Eq. (5.48), we will have:

$$A_n^{(1)}(x) = \int_{-\infty}^{\tau} dz \int_{-\infty}^{z} g_{DBM}(\xi)g^1_{\Theta_n}(z-\xi)d\xi,$$

and also

$$A_n^{(1)}(x) = \int_{-\infty}^{\tau} g_{DBM}(\xi) \left(\int_{\xi}^{\tau} g_{\Theta_n}^1(z-\xi)dz \right) d\xi.$$

By this way, we obtain

$$A_n^{(1)}(x) = \int_{-\infty}^{\tau} g_{DBM}(\xi) \left(\int_{0}^{\tau-\xi} g_{\Theta_n}^1(z)dz \right) d\xi. \tag{5.51}$$

Let us identify the number $r_{j,n}^{(1)}$ in the following way:

$$r_{j,n}^{(1)} := r_{j,n}(p, q, \eta_1, \eta_2). \tag{5.52}$$

where,

$$r_{j,n}(p, q, \eta_1, \eta_2)$$
$$= \begin{cases} \frac{(p\eta_1)^n}{(n-1)!}, & \text{if } j = n-1 \text{ and } n \geq 1, \\ \frac{(n-j-2)!}{(\eta_1+\eta_2)^{n-j-1}} \left(\sum_{k=j+1}^{n-1} \frac{\binom{n}{k}\binom{j}{k-1}(p\eta_1)^k(q\eta_2)^{n-k}}{(k-1)!(n-k-1)!} \right), & \text{if } j = 0, \ldots, n-2 \text{ and } n \geq 2, \end{cases} \tag{5.53}$$

In accordance with Eqs. (5.34), (5.38) and (5.52), we will get:

$$g_{\Theta_n}^1(z) = \sum_{j=0}^{n-1} r_{j,n}^{(1)} z^j e^{-\eta_1 z}. \tag{5.54}$$

Note that for $a \in \mathbb{R}$ and $j = 0, 1, \ldots$, the following representation holds:

$$\int_0^a z^j e^{-\varphi z} dz = \frac{j!}{\varphi^{j+1}} \left[1 - e^{-a\varphi} \sum_{r=0}^{j} \frac{(a\varphi)^r}{r!} \right]. \tag{5.55}$$

From here by using Eq. (5.54) it follows that,

$$\int_0^{\tau-\xi} g_{\Theta_n}^1(\xi)d\xi = \sum_{j=0}^{n-1} \frac{j! r_{j,n}^{(1)}}{\eta_1^{j+1}} \left[1 - e^{-\eta_1(\tau-\xi)} \sum_{r=0}^{j} \frac{(\eta_1(\tau-\xi))^r}{r!} \right],$$

or, which is equivalent also,

$$\int_0^{\tau-\xi} g_{\Theta_n}^1(\xi)d\xi = \sum_{j=0}^{n-1} \frac{j! r_{j,n}^{(1)}}{\eta_1^{j+1}} - e^{-\eta_1(\tau-\xi)} \sum_{r=0}^{n-1} \left[\sum_{j=r}^{n-1} \frac{j! r_{j,n}^{(1)}}{\eta_1^{j+1-r}} \right] \frac{(\tau-\xi)^r}{r!}. \tag{5.56}$$

Taking into account Eqs. (5.51) and (5.56), we will get:

$$A_n^1(x) = \left[\sum_{j=0}^{n-1} \frac{j! r_{j,n}^{(1)}}{\eta_1^{j+1}}\right] \int_{-\infty}^{\tau} g_{DBM}(\xi) d\xi$$
$$- \sum_{r=0}^{n-1} \left[\sum_{j=r}^{n-1} \frac{j! r_{j,n}^{(1)}}{\eta_1^{j+1-r}}\right] \frac{1}{r!} \int_{-\infty}^{\tau} g_{DBM}(\xi)(\tau - \xi)^r e^{-\eta_1(\tau-\xi)} d\xi. \quad (5.57)$$

Let us simplify the expression for $A_n^2(x)$. In accordance with Eqs. (5.49) and (5.50), we shall have

$$A_n^2(x) = \int_{-\infty}^{\tau} dz \int_{-\infty}^{0} g_{DBM}(z-\xi) g_{\Theta_n}^2(\xi) d\xi$$
$$= \int_{-\infty}^{+\infty} dz \int_{-\infty}^{0} g_{DBM}(z-\xi) g_{\Theta_n}^2(\xi) d\xi - \int_{\tau}^{+\infty} dz \int_{-\infty}^{0} g_{DBM}(z-\xi) g_{\Theta_n}^2(\xi) d\xi$$
$$= \int_{-\infty}^{0} g_{\Theta_n}^2(\xi) \left(\int_{-\infty}^{+\infty} g_{DBM}(z-\xi) dz\right) d\xi$$
$$- \int_{\tau}^{+\infty} dz \int_{z}^{+\infty} g_{DBM}(\xi) g_{\Theta_n}^2(z-\xi) d\xi$$
$$= \left[\int_{-\infty}^{+\infty} g_{DBM}(z) dz\right] \left[\int_{-\infty}^{0} g_{\Theta_n}^2(\xi) d\xi\right]$$
$$- \int_{\tau}^{+\infty} g_{DBM}(\xi) \left[\int_{\tau}^{\xi} g_{\Theta_n}^2(z-\xi) dz\right] d\xi,$$

or we will have the following expression, which means the same value:

$$A_n^2(x) = \left[\int_{-\infty}^{+\infty} g_{DBM}(z) dz\right] \left[\int_{-\infty}^{0} g_{\Theta_n}^2(\xi) d\xi\right] - \int_{\tau}^{+\infty} g_{DBM}(\xi) \left[\int_{\tau-\xi}^{0} g_{\Theta_n}^2(z) dz\right] d\xi. \quad (5.58)$$

By using Eq. (5.36), the expression for $g_{\Theta_n}^2(z)$ will be given in the following form:

$$g_{\Theta_n}^2(z) = \sum_{j=0}^{n-1} r_{j,n}^{(2)} (-z)^j e^{\eta_2 z}, \quad (5.59)$$

where,

$$r_{j,n}^{(2)} := r_{j,n}(q, p, \eta_2, \eta_1), \quad (5.60)$$

and $r_{j,n}(q, p, \eta_2, \eta_1)$ is identified by Eq. (5.53).

Taking the above formulas into account, we can rewrite Eq. (5.58) in the following representation, by making the change of variables $\xi = -\xi$ in the minuend, and $z = -z$ in the subtrahend:

$$A_n^2(x) = \left[\int_{-\infty}^{+\infty} g_{DBM}(z)dz\right]\left[\int_0^{\infty} g_{\Theta_n}^2(-\xi)d\xi\right] - \int_{\tau}^{+\infty} g_{DBM}(\xi)\left[\int_0^{\xi-\tau} g_{\Theta_n}^2(-z)dz\right]d\xi. \tag{5.61}$$

If we will use Eqs. (5.55) and (5.59), we will have,

$$\int_0^{\infty} g_{\Theta_n}^2(-\xi)d\xi = \sum_{j=0}^{n-1}\frac{j!r_{j,n}^2}{\eta_2^{j+1}}, \tag{5.62}$$

$$\int_0^{\xi-\tau} g_{\Theta_n}^2(-z)dz = \sum_{j=0}^{n-1}\frac{j!r_{j,n}^2}{\eta_2^{j+1}} - e^{-\eta_2(\xi-\tau)}\sum_{r=0}^{n-1}\left[\sum_{j=r}^{n-1}\frac{j!r_{j,n}^2}{\eta_2^{j+1-r}}\right]\frac{(\xi-\tau)^r}{r!}. \tag{5.63}$$

By combining Eqs. (5.62)–(5.63) in Eq. (5.61), we shall have the subsequent representation for A_n^2:

$$A_n^2(x) = \left[\sum_{j=0}^{n-1}\frac{j!r_{j,n}^2}{\eta_2^{j+1}}\right]\int_{-\infty}^{\tau} g_{DBM}(\xi)d\xi + \sum_{r=0}^{n-1}\left[\sum_{j=r}^{n-1}\frac{j!r_{j,n}^2}{\eta_2^{j+1-r}}\right]\frac{1}{r!}\int_{\tau}^{+\infty} g_{DBM}(\xi)(\xi-\tau)^r e^{-\eta_2(\xi-\tau)}d\xi. \tag{5.64}$$

5.4 Distribution Function of L_t

Let us introduce $\phi(x)$ as the *Laplace function* (or the *Error function*) [1],

$$\phi(x) = \frac{2}{\sqrt{\pi}}\int_0^x e^{-\zeta^2}d\zeta. \tag{5.65}$$

It is well-known that [1],

$$\phi(+\infty) = 1. \tag{5.66}$$

From Eqs. (5.65) and (5.66) we may infer the following:

$$\int_0^x e^{-\zeta^2} d\zeta = \frac{\sqrt{\pi}}{2}\phi(x), \tag{5.67}$$

$$\int_0^{+\infty} e^{-\zeta^2} d\zeta = \frac{\sqrt{\pi}}{2}. \tag{5.68}$$

Lemma 5.1 *The following equalities hold:*

$$\int_0^{+\infty} e^{-\zeta^2}\zeta^{2l} d\zeta = \frac{1}{2}\Gamma(l+\frac{1}{2}) = \frac{(2l)!}{l!}\frac{\sqrt{\pi}}{2^{2l+1}}, \tag{5.69}$$

$$\int_0^{+\infty} e^{-\zeta^2}\zeta^{2l+1} d\zeta = \frac{1}{2}\Gamma(l+1) = \frac{l!}{2}, \tag{5.70}$$

for all $l = 0, 1, 2, \ldots$.

Lemma 5.2 *The following equalities hold:*

$$\int_0^x e^{-\zeta^2}\zeta^{2l+1} d\zeta = \frac{l!}{2} - \frac{l!}{2}\sum_{r=0}^{l}\frac{x^{2r}}{r!}e^{-x^2} \quad (l = 0, 1, 2, \ldots), \tag{5.71}$$

$$\int_0^x e^{-\zeta^2}\zeta^{2l} d\zeta = \frac{(2l)!\sqrt{\pi}}{l!2^{2l+1}}\phi(x) - \frac{(2l)!}{l!2^{2l+1}}\sum_{r=1}^{l}\frac{r!2^{2r}}{(2r)!}x^{2r-1}e^{-x^2} \quad (l = 1, 2, \ldots), \tag{5.72}$$

where $x \in \mathbb{R}$.

Let us assume that $a \in \mathbb{R}$, $b > 0$ and

$$M_j(a, b) = M_j = \int_0^{+\infty} e^{-\frac{(x-a)^2}{b^2}} x^j dx \quad (j = 0, 1, 2, \ldots). \tag{5.73}$$

In the following steps, we shall get the expression of M_j in terms of $\phi(x)$.

If we make a change of $x = a + b\zeta$ in Eq. (5.73), then we will have

$$M_j = b\int_{-\frac{a}{b}}^{+\infty} e^{-\zeta^2}(a + b\zeta)^j d\zeta. \tag{5.74}$$

By assuming that $\theta = a/b$, and using Eq. (5.74), we will get the following representation:

$$M_j = b^{j+1}\int_{-\theta}^{+\infty} e^{-\zeta^2}(\theta + \zeta)^j d\zeta. \tag{5.75}$$

For $j = 0$, Eq. (5.74) obtains a form as follows:

$$M_0 = b \int_{-\theta}^{+\infty} e^{-\zeta^2} d\zeta; \tag{5.76}$$

and by applying Eqs. (5.67) and (5.68) to the above expression we shall get:

$$\int_{-\theta}^{+\infty} e^{-\zeta^2} d\zeta = \int_{0}^{+\infty} e^{-\zeta^2} d\zeta + \int_{0}^{\theta} e^{-\zeta^2} d\zeta = \frac{\sqrt{\pi}}{2} + \frac{\sqrt{\pi}}{2}\phi(\theta). \tag{5.77}$$

Hence, Eq. (5.76) will be given in the subsequent representation,

$$M_0 = \frac{b\sqrt{\pi}}{2} + \frac{b\sqrt{\pi}}{2}\phi(\theta). \tag{5.78}$$

Let us analyze the cases of $j \geq 1$. Equation (5.75) can be simplified in the following manner:

$$\begin{aligned} M_j &= b^{j+1} \sum_{r=0}^{j} \binom{j}{r} \theta^{j-r} \int_{-\theta}^{+\infty} e^{-\zeta^2} \zeta^r d\zeta \\ &= b^{j+1} \sum_{l=0}^{[\frac{j}{2}]} \binom{j}{2l} \theta^{j-2l} \int_{-\theta}^{+\infty} e^{-\zeta^2} \zeta^{2l} d\zeta \\ &+ b^{j+1} \sum_{l=0}^{[\frac{j-1}{2}]} \binom{j}{2l+1} \theta^{j-2l-1} \int_{-\theta}^{+\infty} e^{-\zeta^2} \zeta^{2l+1} d\zeta, \end{aligned} \tag{5.79}$$

where $[c]$ simply depicts the integer part of the c.

Note that for $\nu = 0, 1, \ldots,$ the next equation holds:

$$\int_{-\theta}^{0} e^{-\zeta^2} \zeta^{\nu} d\zeta = (-1)^{\nu} \int_{0}^{\theta} e^{-\zeta^2} \zeta^{\nu} d\zeta. \tag{5.80}$$

Hence,

$$\int_{-\theta}^{+\infty} e^{-\zeta^2} \zeta^{\nu} d\zeta = \int_{0}^{+\infty} e^{-\zeta^2} \zeta^{\nu} d\zeta + (-1)^{\nu} \int_{0}^{\theta} e^{-\zeta^2} \zeta^{\nu} d\zeta. \tag{5.81}$$

From Eqs. (5.69)–(5.72) and (5.81), we can infer the subsequent simplification:

$$\int_{-\theta}^{+\infty} e^{-\zeta^2}\zeta^{2l}d\zeta = \frac{(2l)!}{l!2^{2l+1}}\left[\sqrt{\pi} + \sqrt{\pi}\phi(\theta) - e^{-\theta^2}\sum_{r=1}^{l}\frac{r!2^{2r}}{(2r)!}\theta^{2r-1}\right] (l = 1, 2, \ldots), \tag{5.82}$$

and

$$\int_{-\theta}^{+\infty} e^{-\zeta^2}\zeta^{2l+1}d\zeta = \frac{l!}{2}e^{-\theta^2}\sum_{r=0}^{l}\frac{\theta^{2r}}{r!} \quad (l = 0, 1, 2, \ldots). \tag{5.83}$$

By taking into account Eqs. (5.77), (5.79) and (5.83), we will have the following formula for M_1:

$$M_1 = \frac{b^2}{2}\left(\theta\sqrt{\pi} + \theta\sqrt{\pi}\phi(\theta) + e^{-\theta^2}\right). \tag{5.84}$$

Equation (5.79) can be written as stated below for the cases of $j \geq 2$:

$$\begin{aligned} M_j =& b^{j+1}\theta^j\int_{-\theta}^{+\infty} e^{-\zeta^2}d\zeta + b^{j+1}\sum_{l=1}^{[\frac{j}{2}]}\binom{j}{2l}\theta^{j-2l}\int_{-\theta}^{+\infty} e^{-\zeta^2}\zeta^{2l}d\zeta + \\ &+ b^{j+1}\sum_{l=0}^{[\frac{j-1}{2}]}\binom{j}{2l+1}\theta^{j-2l-1}\int_{-\theta}^{+\infty} e^{-\zeta^2}\zeta^{2l+1}d\zeta. \end{aligned} \tag{5.85}$$

By taking into account Eqs. (5.77), (5.82) and (5.83), we will get the following expression for Eq. (5.85) in cases of $j \geq 2$:

$$\begin{aligned} M_j =& \frac{b^{j+1}}{2}\left(\sqrt{\pi}\sum_{l=0}^{[j/2]}\binom{j}{2l}\frac{(2l)!}{l!2^{2l}}\theta^{j-2l} + \sqrt{\pi}\phi(\theta)\sum_{l=0}^{[\frac{j}{2}]}\binom{j}{2l}\frac{(2l)!}{l!2^{2l}}\theta^{j-2l} + \right. \\ &+ e^{-\theta^2}\sum_{l=0}^{[\frac{j-1}{2}]} l!\binom{j}{2l+1}\sum_{r=0}^{l}\frac{\theta^{2r+j-2l-1}}{r!} \\ &\left. - e^{-\theta^2}\sum_{l=1}^{[\frac{j}{2}]}\binom{j}{2l}\frac{(2l)!}{l!2^{2l}}\sum_{r=1}^{l}\frac{r!2^{2r}}{(2r)!}\theta^{2r+j-2l-1}\right). \end{aligned} \tag{5.86}$$

In can be easily inferred that,

$$\begin{aligned}\sum_{l=0}^{[\frac{j-1}{2}]} l!\binom{j}{2l+1}\sum_{r=0}^{l}\frac{\theta^{j-1-2(l-r)}}{r!} &= \sum_{l=0}^{[\frac{j-1}{2}]} l!\binom{j}{2l+1}\sum_{r=0}^{l}\frac{\theta^{j-1-2r}}{(l-r)!}\\ &= \sum_{r=0}^{[\frac{j-1}{2}]}\theta^{j-1-2r}\sum_{l=r}^{[(j-1)/2]}\frac{l!\binom{j}{2l+1}}{(l-r)!},\end{aligned}$$

and

$$\begin{aligned}\sum_{l=1}^{[\frac{j}{2}]}\frac{(2l)!\binom{j}{2l}}{l!2^{2l}}\sum_{r=1}^{l}\frac{r!2^{2r}}{(2r)!}\theta^{j-2(l-r)-1} &= \sum_{l=1}^{[\frac{j}{2}]}\frac{(2l)!\binom{j}{2l}}{l!2^{2l}}\sum_{r=0}^{l-1}\frac{(l-r)!2^{2(l-r)}}{(2(l-r))!}\theta^{j-2r-1}\\ &= \sum_{r=0}^{[\frac{j}{2}]-1}\theta^{j-2r-1}\sum_{l=r+1}^{[j/2]}\frac{(2l)!(l-r)!\binom{j}{2l}}{l!(2(l-r))!2^{2r}}.\end{aligned}$$

From the above expressions and from Eq. (5.86), for $j \geq 2$, the following is true:

$$\begin{aligned}M_j =& \frac{\sqrt{\pi}}{2}b^{j+1}\left[1+\phi(\theta)\right]\sum_{l=0}^{[\frac{j}{2}]}\alpha_{l,j}\theta^{j-2l}+\\ &+\frac{b^{j+1}}{2}e^{-\theta^2}\left(\sum_{l=0}^{[\frac{j-1}{2}]}\beta_{l,j}\theta^{j-2l-1}+\sum_{l=1}^{[\frac{j}{2}]-1}\gamma_{l,j}\theta^{j-2l-1}\right),\end{aligned}\tag{5.87}$$

where,

$$\alpha_{l,j} := \frac{(2l)!}{l!2^{2l}}\binom{2l}{j},\tag{5.88}$$

$$\beta_{l,j} := \sum_{r=l}^{[\frac{j-1}{2}]}\frac{r!}{(r-l)!}\binom{2r+1}{j},\tag{5.89}$$

and

$$\gamma_{l,j} := -\frac{1}{2^{2l}}\sum_{r=l+1}^{[\frac{j}{2}]}\frac{(2r)!(r-l)!}{r!(2(r-l))!}\binom{2r}{j}.\tag{5.90}$$

Below let us introduce some denotation:

$$\sigma_0 := (\mu - \frac{\sigma^2}{2})t + \ln(s_0), \quad \theta_0 := \frac{\tau - \sigma_0}{\sigma\sqrt{2t}}. \tag{5.91}$$

By using the above designations and Eq. (5.42), distribution function of Drift plus Brownian Motion, we will get the subsequent representation for the integral part of the minuend of Eq. (5.57):

$$\int_{-\infty}^{\tau} g_{DBM}(\xi)d\xi = \frac{1}{\sigma\sqrt{2\pi t}} \int_{-\infty}^{\tau} e^{-\frac{(\xi-\sigma_0)^2}{2\sigma^2 t}} d\xi.$$

Making the variable change of

$$\frac{\xi - \sigma_0}{\sigma\sqrt{2t}} = u, \tag{5.92}$$

and using the variables described in Eq. (5.91) will result in the following expression:

$$\begin{aligned}\int_{-\infty}^{\tau} g_{DBM}(\xi)d\xi &= \frac{1}{\sqrt{\pi}} \int_{-\infty}^{\frac{\tau-\sigma_0}{\sigma\sqrt{2t}}} e^{-u^2} du \\ &= \frac{1}{\sqrt{\pi}} \int_{-\infty}^{0} e^{-u^2} du + \frac{1}{\sqrt{\pi}} \int_{0}^{\theta_0} e^{-u^2} du.\end{aligned}$$

From the above integral and in accordance with Eqs. (5.67) and (5.68), the subsequent will follow:

$$\int_{-\infty}^{\tau} g_{DBM}(\xi)d\xi = \frac{1}{2} + \frac{1}{2}\phi(\theta_0). \tag{5.93}$$

Let us introduce the below designations:

$$\sigma_1 := \tau - \sigma_0 - \sigma^2 t\eta_1, \tag{5.94}$$

$$T_1 := e^{-\eta_1(\tau-\sigma_0-\frac{1}{2}\sigma^2 t\eta_1)}. \tag{5.95}$$

After a simple transformation, the integral in the subtrahend of Eq. (5.57) will obey the following formula:

$$\begin{aligned}\int_{-\infty}^{\tau} g_{DBM}(\xi)(\tau-\xi)^r e^{-\eta_1(\tau-\xi)} d\xi &= \int_{0}^{\infty} g_{DBM}(\tau-\xi)\xi^r e^{-\eta_1\xi} d\xi \\ &= \frac{T_1}{\sigma\sqrt{2\pi t}} \int_{0}^{\infty} e^{-\frac{(\xi-\sigma_1)^2}{2\sigma^2 t}} \xi^r d\xi,\end{aligned} \tag{5.96}$$

where σ_1 and T_1 are identified by the equalities of Eqs. (5.94) and (5.95) accordingly. Analogically, for the integral in subtrahend of Eq. (5.64) we will have,

$$\begin{aligned}\int_{\tau}^{\infty} g_{DBM}(\xi)(\xi-\tau)^r e^{-\eta_2(\xi-\tau)}d\xi &= \int_{0}^{\infty} g_{DBM}(\xi+\tau)\xi^r e^{-\eta_2\xi}d\xi \\ &= \frac{T_2}{\sigma\sqrt{2\pi t}}\int_0^{\infty} e^{-\frac{(\xi-\sigma_2)^2}{2\sigma^2 t}}\xi^r d\xi,\end{aligned} \tag{5.97}$$

where,

$$\sigma_2 := \sigma_0 - \tau - \sigma^2 t\eta_2, \tag{5.98}$$

$$T_2 := e^{-\eta_2\left(\sigma_0-\tau-\frac{1}{2}\sigma^2 t\eta_2\right)}. \tag{5.99}$$

Equations (5.96) and (5.97) can be represented by using Eq. (5.73) as stated below:

$$\int_{-\infty}^{\tau} g_{DBM}(\xi)(\tau-\xi)^r e^{-\eta_1(\tau-\xi)}d\xi = \frac{T_1}{\sigma\sqrt{2\pi t}}M_r(\sigma_1,\sigma\sqrt{2t}), \tag{5.100}$$

$$\int_{\tau}^{\infty} g_{DBM}(\xi)(\xi-\tau)^r e^{-\eta_2(\xi-\tau)}d\xi = \frac{T_2}{\sigma\sqrt{2\pi t}}M_r(\sigma_2,\sigma\sqrt{2t}). \tag{5.101}$$

Let us introduce $D_{r,n}^{(s)}$ by:

$$D_{r,n}^{(s)} := \frac{1}{r!}\sum_{j=r}^{n-1}\frac{j!r_{j,n}^{(s)}}{\eta_s^{j+1-r}}. \tag{5.102}$$

From Eqs. (5.57), (5.64), (5.93) and (5.100)–(5.102), we will have

$$A_n^{(s)}(x) = \frac{1}{2}D_{0,n}^{(s)}(1+\phi(\theta_0)) + \frac{(-1)^s T_s}{\sigma\sqrt{2\pi t}}\sum_{r=0}^{n-1}D_{r,n}^{(s)}M_r(\sigma_s,\sigma\sqrt{2t}), \tag{5.103}$$

where $s = 1, 2$.

In accordance with Eq. (5.103),

$$\begin{aligned}\sum_{n=1}^{\infty}A_n^{(s)}(x)\frac{(\lambda t)^n}{n!}e^{-\lambda t} =& \frac{1}{2}(1+\phi(\theta_0))\sum_{n=1}^{\infty}D_{0,n}^{(s)}\frac{(\lambda t)^n}{n!}e^{-\lambda t} \\ &+ \frac{(-1)^s T_s}{\sigma\sqrt{2\pi t}}\sum_{n=1}^{\infty}\left[\sum_{r=0}^{n-1}D_{r,n}^{(s)}M_r(\sigma_s,\sigma\sqrt{2t})\right]\frac{(\lambda t)^n}{n!}e^{-\lambda t}.\end{aligned}$$

The above formula can be revised in the following way:

$$\sum_{n=1}^{\infty} A_n^{(s)}(x)\frac{(\lambda t)^n}{n!}e^{-\lambda t} = \frac{e^{-\lambda t}}{2}(1+\phi(\theta_0))\sum_{n=1}^{\infty} D_{0,n}^{(s)}\frac{(\lambda t)^n}{n!}$$
$$+\frac{(-1)^s T_s e^{-\lambda t}}{\sigma\sqrt{2\pi t}}\sum_{r=0}^{\infty}\left[\sum_{n=r+1}^{n-1} D_{r,n}^{(s)}\frac{(\lambda t)^n}{n!}\right] M_r(\sigma_s, \sigma\sqrt{2t}). \tag{5.104}$$

Let us introduce,

$$\theta_s := \frac{\sigma_s}{\sigma\sqrt{2t}} \quad (s=1,2), \tag{5.105}$$

and

$$B_r^{(s)} := \sum_{n=r+1}^{\infty} D_{r,n}^{(s)}\frac{(\lambda t)^n}{n!}e^{-\lambda t}. \tag{5.106}$$

By using the variables identified above, Eq. (5.104) will have the representation below:

$$\sum_{n=1}^{\infty} A_n^{(s)}(x)\frac{(\lambda t)^n}{n!}e^{-\lambda t} = \frac{B_0^{(s)}}{2}(1+\phi(\theta_0)) + \frac{(-1)^s T_s}{\sigma\sqrt{2\pi t}}\sum_{r=0}^{\infty} B_r^{(s)} M_r(\sigma_s, \sigma\sqrt{2t}). \tag{5.107}$$

From Eqs. (5.78), (5.84) and (5.87), we can infer the following:

$$M_0(\sigma_s, \sigma\sqrt{2t}) = \frac{\sigma\sqrt{2\pi t}}{2}(1+\phi(\theta_s)), \tag{5.108}$$

$$M_1(\sigma_s, \sigma\sqrt{2t}) = \frac{(\sigma\sqrt{2t})^2}{2}(\theta_s\sqrt{\pi} + \theta_s\sqrt{\pi}\phi(\theta_s) + e^{\theta_s^2}), \tag{5.109}$$

$$M_r(\sigma_s, \sigma\sqrt{2t}) = \frac{\sqrt{\pi}}{2}(\sigma\sqrt{2t})^{r+1}(1+\phi(\theta_s))\sum_{l=0}^{[\frac{r}{2}]}\alpha_{l,r}\theta_s^{r-2l}$$
$$+\frac{1}{2}(\sigma\sqrt{2t})^{r+1}e^{-\theta_s^2}\left[\sum_{l=0}^{[\frac{r-1}{2}]}\beta_{l,r}\theta_s^{r-2l-1} + \sum_{l=0}^{[\frac{r}{2}]-1}\gamma_{l,r}\theta_s^{r-2l-1}\right], \tag{5.110}$$

where $r \geq 2$.

By using Eqs. (5.107)–(5.110) and making some simple transformations, we will get this equality:

$$\begin{aligned}\sum_{n=1}^{\infty} A_n^{(s)}(x)\frac{(\lambda t)^n}{n!}e^{-\lambda t} =& \frac{B_0^{(s)}}{2}(1+\phi(\theta_0))\\ &+\frac{(-1)^s T_s}{2}(1+\phi(\theta_s))\sum_{r=0}^{\infty} B_r^{(s)}(\sigma\sqrt{2t})^r\sum_{l=0}^{[\frac{r}{2}]}\alpha_{l,r}\theta_s^{r-2l}\\ &+\frac{(-1)^s T_s}{2}e^{-\theta_s^2}\Bigg[\sum_{r=1}^{\infty} B_r^{(s)}(\sigma\sqrt{2t})^r\sum_{l=0}^{[\frac{r-1}{2}]}\beta_{l,r}\theta_s^{r-2l-1}\\ &+\sum_{r=2}^{\infty} B_r^{(s)}(\sigma\sqrt{2t})^r\sum_{l=0}^{[\frac{r}{2}]-1}\gamma_{l,r}\theta_s^{r-2l-1}\Bigg],\end{aligned} \tag{5.111}$$

where

$$\alpha_{0,0} = \beta_{0,1} = 1. \tag{5.112}$$

Equation (5.111) can be revised in the following way:

$$\begin{aligned}\sum_{n=1}^{\infty} A_n^{(s)}(x)\frac{(\lambda t)^n}{n!}e^{-\lambda t} =& \frac{B_0^{(s)}}{2}(1+\phi(\theta_0))\\ &+(1+\phi(\theta_s))\sum_{m=0}^{\infty} d_m^{(s)}\theta_s^m + e^{-\theta_s^2}\sum_{m=0}^{\infty} D_m^{(s)}\theta_s^m,\end{aligned} \tag{5.113}$$

where

$$d_m^{(s)} := \begin{cases} \frac{(-1)^s T_s}{2}\sum_{r=l}^{\infty} B_{2r}^{(s)}(\sigma\sqrt{2t})^{2r}\alpha_{r-l,2r}, & \text{if} \quad m=2l,\\ \frac{(-1)^s T_s}{2}\sum_{r=l}^{\infty} B_{2r+1}^{(s)}(\sigma\sqrt{2t})^{2r+1}\alpha_{r-l,2r+1}, & \text{if} \quad m=2l+1,\end{cases} \tag{5.114}$$

and

$$D_m^{(s)} := \begin{cases} \frac{(-1)^s T_s}{2\sqrt{\pi}}\sum_{r=0}^{\infty} B_{2r+1}^{(s)}(\sigma\sqrt{2t})^{2r+1}\beta_{r,2r+1}, & \text{if} \quad m=0,\\ \frac{(-1)^s T_s}{2\sqrt{\pi}}\sum_{r=l}^{\infty} B_{2r+1}^{(s)}(\sigma\sqrt{2t})^{2r+1}(\beta_{r-l,2r+1}+\gamma_{r-l,2r+1}), & \text{if} \quad m=2l \quad (l=1,2,\ldots),\\ \frac{(-1)^s T_s}{2\sqrt{\pi}}\sum_{r=l+1}^{\infty} B_{2r}^{(s)}(\sigma\sqrt{2t})^{2r}(\beta_{r-l-1,2r}+\gamma_{r-l-1,2r}), & \text{if} \quad m=2l+1 \quad (l=0,1,\ldots).\end{cases} \tag{5.115}$$

Finally, from Eqs. (5.47) and (5.113) we will get the subsequent distribution function for S_t $(t \geq 0)$:

$$F_{S_t}(x) = \frac{1}{2}\left(B_0^{(1)} + B_0^{(2)}\right)(1 + \phi(\theta_0)) + (1 + \phi(\theta_1)) \sum_{m=0}^{\infty} d_m^{(1)} \theta_1^m + (1 + \phi(\theta_2)) \sum_{m=0}^{\infty} d_m^{(2)} \theta_2^m + e^{-\theta_1^2} \sum_{m=0}^{\infty} D_m^{(1)} \theta_1^m + e^{-\theta_2^2} \sum_{m=0}^{\infty} D_m^{(2)} \theta_2^m. \tag{5.116}$$

5.5 Risk-Neutral Dynamics

Under a risk-neutral measure, the financial instrument must grow at a risk free rate in order for the discounted prices to be a martingale. Hence, the very natural thing to do is to specify the mean return rate, μ, that will be equal to a risk free rate, which have been done for the Black-Scholes case and moreover refer to other variables as a risk neutral, too. For any model that incorporates jump, as in our case the jump-diffusion model, there are more that one equivalent risk-neutral measures $\mathbb{Q} \sim \mathbb{P}$ where the discounted prices become martingales. In another words, there is no way an investor would be able to construct a hedging portfolio that is totally risk-free.

Incompleteness of the market is one of the crucial problems that we are facing with jump-diffusion or Lévy processes. Surely, the risk arousing from jumps is unhedgeable and no exclusive risk-neutral measure exists. Researchers suggest that several measures with distant characteristics can be selected, such as Hellinger distance, L^2 distance, entropy or Kullback-Leibler distance, which would be equivalent to the historical probability measure.

We select the *Esscher measure* as the change of measure technique that is based on the compound return of the price process. Hans U. Gerber in his article [12] has developed this change of measure. Our choice relies on several reasons. Firstly, a mathematical viewpoint, in the sense of power metrics Esscher measure can be regarded as the equivalent martingale measure closest to the real-world probability measure. Secondly, we refer to an economic point of view. As the risk-neutral universe for jump-diffusion models is not unique, the behavior of economic agents toward risk will shape the prices. Thirdly, by employment of either neo-Bernoulli theory or expected utility, we can achieve "clean" price with the principle of marginal utility. In fact, it could be demonstrated that the proper prices derived from a logarithmic utility or power utility function can be formulated by using the Esscher measure that employs compound return calculation.

Asset value can be given by the exponent of Lévy process X or by the Doléans-Dade exponential of some other process $\tilde{X}$. The processes could be connected among each other and connections between the characteristic triplets may be quickly calculated. Hence, we are allowed to speak about the Esscher measure, specified with $\tilde{X}$. From the economic viewpoint, the given measure is associated with the utility functions of exponential forms. Furthermore, when the given measure is applied, Lévy processes stay as Lévy processes during the transition from the real-world universe to the new risk-neutral universe. Under this risk measure, the characteristic triplet of the process X also has a ready to use common formula.

Let us now employ of the Esscher transform in the framework of our universe. Following Radon-Nikodym formula characterizes the Esscher risk-neutral measure with one parameter, indicated by $\mathbb{Q}$, which is related with the process X and the parameter u:

$$\left(\frac{d\mathbb{Q}}{d\mathbb{P}}\right)_t = \frac{e^{uX_t}}{\mathbb{E}_{\mathbb{P}}\left[e^{uX_t}\right]}.$$

The risk-neutral measure on the Laplace exponent $\psi_{\mathbb{Q}}(\beta)$ can be represented by the next formula:

$$\mathbb{E}_{\mathbb{Q}}\left[e^{\beta X_t}\right] = \mathbb{E}_{\mathbb{P}}\left[\eta_t e^{\beta X_t}\right] = e^{t\psi_{\mathbb{Q}}(\beta)}.$$

Let X be a Lévy process, and $\mathbb{Q}_u$ be a u-Esscher measure related to X and specified by:

$$\frac{d\mathbb{Q}_u}{d\mathbb{P}} = \frac{e^{uX_t}}{\mathbb{E}_{\mathbb{P}}\left[e^{uX_t}\right]}.$$

The Laplace exponent of $\psi(\beta)$ would be such that,

$$\mathbb{E}\left(e^{\beta X_t}\right) = e^{t\psi(\beta)}.$$

Let us first consider the martingale condition that under in risk neutral world the discounted process must be a martingale. For $t \geq 0$ this condition can be interpreted in the subsequent way:

$$S_t = \mathbb{E}_{\mathbb{Q}_u}\left(S_0 e^{X(t)} e^{-rt}\right),$$

or

$$S_t = S_0 e^{-rt} \int_{\Omega} \frac{e^{(u+1)X(t)}}{\mathbb{E}_{\mathbb{P}}(e^{uX(t)})} d\mathbb{P}.$$

This latter condition writes as

$$S_t = S_0 e^{-rt} \frac{\psi_{\mathbb{P}}(u+1)}{\psi_{\mathbb{P}}(u)},$$

where real-world probability is designated by $\psi_{\mathbb{P}}$. Lastly, the martingale condition conforms the description of an answer of the next equation which is identified by the parameter u^*:

$$-r + \psi_{\mathbb{P}(u+1)} - \psi_{\mathbb{P}(u)} = 0. \tag{5.117}$$

The main purpose of this part of the thesis is to identify the link between parameters of the Kou model in the historical and in the risk-neutral world, by using the rationale of the Laplace exponent under $\mathbb{Q}_{u^*}$. We will designate the solution of Eq. (5.117) by u instead of u^*, and by $\mathbb{Q}$ in lieu of $\mathbb{Q}_{u^*}$ the risk-neutral measure. In this regard, we can easily have:

$$\mathbb{E}_{\mathbb{Q}}\left(e^{\beta X(t)}\right) = \int_{\Omega} e^{\beta X(t)} \frac{e^{uX(t)}}{\mathbb{E}_{\mathbb{P}}(e^{uX(t)})} d\mathbb{P},$$

or

$$e^{\psi_{\mathbb{Q}}(\beta)} = \mathbb{E}_{\mathbb{P}}\left(\frac{e^{(\beta+u)X(t)}}{\mathbb{E}_{\mathbb{P}}(e^{uX(t)})}\right) = \frac{e^{\psi_{\mathbb{P}}(\beta+u)}}{e^{\psi_{\mathbb{P}}(u)}}.$$

This generates a very easy equation, associating the Laplace exponent in the (Esscher) risk-neutral measure with the same exponent in the historical universe:

$$\psi_{\mathbb{Q}}(\beta) = \psi_{\mathbb{P}}(\beta + u) - \psi_{\mathbb{P}}(u). \tag{5.118}$$

Let us employ the above general outcome in the framework of Kou model, which has the subsequent expression:

$$X(t) = \left(\mu - \frac{1}{2}\sigma^2\right) t + \sigma W_t + \sum_{k=1}^{N_t} Y_k \quad (t \geq 0).$$

Kou process has the subsequent Laplace exponent

$$\psi_{\mathbb{P}}(\beta) = \frac{1}{2}\sigma^2\beta^2 + \left(\mu - \frac{1}{2}\sigma^2\right)\beta + \lambda\left[\frac{p\eta_1}{\eta_2 - \beta} + \frac{p\eta_2}{\eta_2 + \beta} - 1\right]. \tag{5.119}$$

Putting the final expression into the martingale condition of Eq. (5.117) yields

$$r = u\sigma^2 + \mu - \lambda\left[\frac{p\eta_1}{(\eta_1 - u)(\eta_1 - (1+u))} - \frac{q\eta_2}{(\eta_2 + u)(\eta_2 + (1+u))}\right].$$

To achieve the corresponding martingale measure, the parameter u is selected in a way so that the discounted price process would be a martingale. It can be easily shown that the risk-neutral parameter u under the Esscher measure can be represented by the below given martingale condition:

$$r = u\sigma^2 + \mu - \lambda \left[\frac{p\eta_1}{\eta_1 - u} + \frac{q\eta_2}{\eta_2 + u} - \frac{p\eta_1}{\eta_1 - (1+u)} - \frac{q\eta_2}{\eta_2 + (1+u)} \right]. \quad (5.120)$$

The Esscher transform, parametrized only over u, is therefore entirely characterized at this stage.

Now, using Eq. (5.119) inside Eq. (5.118), and after some mild computations, one obtains:

$$\psi_{\mathbb{P}}(\beta) = \frac{1}{2}\sigma^2\beta^2 + \left(\mu - \frac{1}{2}\sigma^2 + \sigma^2 u\right)\beta + \hat{\lambda}\left[\frac{\hat{p}\hat{\eta}_1}{\hat{\eta}_2 - \beta} + \frac{\hat{p}\hat{\eta}_2}{\hat{\eta}_2 + \beta} - 1\right], \quad (5.121)$$

where:

$$\begin{cases} \hat{p} := \frac{p\eta_1}{\zeta(\eta_1 - u)}, \\ \hat{q} := 1 - \hat{p}, \\ \hat{\eta}_1 := \eta_1 - u, \\ \hat{\eta}_2 := \eta_2 + u, \\ \hat{\lambda} := \lambda\zeta, \quad \text{and} \\ \zeta := \frac{p\eta_1}{\eta_1 - u} + \frac{q\eta_2}{\eta_2 + u}, \end{cases} \quad (5.122)$$

which is the relationship between the real-world and the risk-neutral parameters.

Equation (5.121) indicates that X is a Kou process under the Esscher risk-neutral measure. The parameters of the process are presented in Eq. (5.122), where $u = u^*$ is the solution of Eq. (5.120). We have to take into account that the characteristic triplet of the Kou process X under the Esscher measure could be achieved directly from the representations of Eq. (5.121).

After the computations exhibited above, we will get the subsequent expression:

$$\psi_{\mathbb{Q}}(\beta) = A\beta + \frac{1}{2}\Gamma^2\beta^2 + \int_{\mathbb{R}} (e^{\beta y} - 1)\hat{\nu}(dy),$$

with $(A, \Gamma, \hat{\nu})$ being the characteristic triplet of the Kou process under the u-Esscher measure, or:

$$A := a + \sigma^2 u, \quad \Gamma := \sigma, \quad \hat{\nu}(dy) := e^{uy}\nu(dy).$$

So, in conclusion, we can say that the gain process would still be a Kou process in the risk-neutral world, although with distant parameters.

We can get all the details of the risk-neutral parameters by using the formulas given by Eq. (5.122). The main conclusion may be summarized in the following way: When the gain process follows a Kou model in the real-world universe, it will follow similar Kou model in the risk-neutral universe, related to the selection of correct Esscher measure. Moreover, under these circumstances the change of parameter can be defined as given in Eq. (5.122) above. Certainly, the formulas in Eq. (5.122) show the way how the coefficients of the real and the risk-neutral world are exactly related. As a summary we may indicate that, in the selected risk-neutral world, the following will hold true:

$$X_t = \left(r - \frac{1}{2}\sigma^2 - \hat{\lambda}\hat{\xi}\right)t + \sigma\hat{W}_t + \sum_{k=1}^{\hat{N}_t} \hat{Y}_k, \tag{5.123}$$

with $\hat{W}$ being a standard $\mathbb{Q}$-Brownian motion, $\hat{N}$ being a Poisson process with fixed intensity rate $\hat{\lambda}$, and the variables $\hat{Y}_k$ being positive i.i.d. random variables of DED with density:

$$\hat{f}_Y(y) = \hat{p}\hat{\eta}_1 e^{-\hat{\eta}_1 y}\mathbb{I}_{\{y\geq 0\}} + \hat{q}\hat{\eta}_2 e^{\hat{\eta}_2 y}\mathbb{I}_{\{y<0\}}, \tag{5.124}$$

under the trivial conditions of $\hat{\eta}_1 > 1$, $\hat{\eta}_2 > 0$, $\hat{p}, \hat{q} > 0$, and $\hat{p} + \hat{q} = 1$.

We have to take into account that $\hat{\xi}$ is identified by $\hat{\xi} = \mathbb{E}_h\left[e^{\hat{Y}_1}\right] - 1 = \hat{p}\frac{\hat{\eta}_1}{\hat{\eta}_1 - 1} + \hat{q}\frac{\hat{\eta}_2}{\hat{\eta}_2 + 1} - 1$, and that all stochastic processes: $\hat{N}$, $\hat{W}$ and the $\hat{Y}_k$, are taken to be independent.

Hence, in the risk-neutral world, asset prices will have the following path:

$$\frac{dS_t}{dS_{t\_}} = rdt + \sigma d\hat{W}_t + d\hat{M}_t, \tag{5.125}$$

where $\hat{M}$ is the compensated martingale given by

$$\hat{M}_t = \sum_{k=1}^{\hat{N}_t} (\hat{Z}_k - 1) - \hat{\lambda}\hat{\xi}t,$$

with $\hat{Z}_k = e^{\hat{Y}_k}$.

Therefore, as a consequence of Itô's Lemma [6], we may indicate that:

$$S_t = S_0 e^{\hat{a} + \sigma\hat{W}_t} \prod_{k=1}^{N_t} \hat{Z}_t,$$

where $\hat{a} = r - \frac{\sigma^2}{2} - \hat{\lambda}\hat{\Phi}(1)$, and where the function $\hat{\Phi}$ is characterized by

$$\hat{\Phi}(x) = \hat{p}\frac{\hat{\eta}_1}{\hat{\eta}_1 - x} + \hat{q}\frac{\hat{\eta}_2}{\hat{\eta}_2 + x} - 1,$$

with $\hat{\Phi}(1) = \hat{\xi}$.

We have completely clarified how a stock price process, designed by using Kou model, is adjusted when changing the historical universe to the Esscher risk-neutral universe.

References

1. G.E. Andrews, R. Askey, R. Roy, *Special Functions*. Encyclopedia of Mathematics and Its Applications. Cambridge University Press, Cambridge (1999). ISBN 9780521789882
2. O.E. Barndorff-Nielsen, N. Shephard, Non-Gaussian Ornstein–Uhlenbeck-based models and some of their uses in financial economics. J. R. Stat. Soc. Ser. B (Stat. Methodol.) **63**(2), 167–241 (2001)
3. R.C. Blattberg, N.J. Gonedes, A comparison of the stable and student distributions as statistical models for stock prices. J. Bus. **47**(2), 244–280 (1974)
4. P. Carr, H. Geman, D.B. Madan, M. Yor, Stochastic volatility for lévy processes. Math. Financ. **13**(3), 345–382 (2003)
5. P.K. Clark, A subordinated stochastic process model with finite variance for speculative prices. Econometrica **141**, 135–155 (1973)
6. R. Cont, P. Tankov, *Financial Modelling with Jump Processes* (Chapman & Hall, Boca Raton, 2004)
7. J.C. Cox, S.A. Ross, The valuation of options for alternative stochastic processes. J. Financ. Econ. **3**(1–2), 145–166 (1976)
8. D. Davydov, V. Linetsky, Pricing and hedging path-dependent options under the CEV process. Manag. Sci. **47**(7), 949–965 (2001)
9. R. Engle, *ARCH: Selected Readings* (Oxford University Press, Oxford, 1995)
10. R. Feldman, M. Taqqu, *A Practical Guide to Heavy Tails: Statistical Techniques and Applications* (Springer Science & Business Media, Berlin, 1998)
11. J.-P. Fouque, G. Papanicolaou, K.R. Sircar, *Derivatives in Financial Markets with Stochastic Volatility* (Cambridge University Press, Cambridge, 2000)
12. H.U. Gerber, E.S. Shiu et al., Option pricing by Esscher transforms. Trans. Soc. Actuaries **46**(99), 140 (1994)
13. S.L. Heston, A closed-form solution for options with stochastic volatility with applications to bond and currency options. Rev. Financ. Stud. **6**(2), 327–343 (1993)
14. C.C. Heyde, A risky asset model with strong dependence through fractal activity time. J. Appl. Probab. **36**, 1234–1239 (1999)
15. C. Heyde, S. Kou, X. Peng, What is a good risk measure: bridging the gaps between robustness, subadditivity, prospect theory, and insurance risk measures. Preprint (2006)
16. J. Hull, A. White, The pricing of options on assets with stochastic volatilities. J. Financ. **42**(2), 281–300 (1987)
17. N.L. Johnson, S. Kotz, N. Balakrishnan, *Continuous Univariate Distribution*, vol. 2 (Wiley, New York, 1995)
18. A.N. Kolmogorov, *Foundations of the Theory of Probability* (Chelsea Publishing Co, New York, 1950)
19. S.G. Kou, A jump-diffusion model for option pricing. Manag. Sci. **48**(8), 1086–1101 (2002)
20. D.B. Madan, E. Seneta, The variance gamma (VG) model for share market returns. J. Bus. **63**, 511–524 (1990)
21. D.B. Madan, P.P. Carr, E.C. Chang, The variance gamma process and option pricing. Eur. Finan. Rev. **2**(1), 79–105 (1998)
22. B.B. Mandelbrot, The variation of certain speculative prices, in *Fractals and Scaling in Finance* (Springer, Berlin, 1997), pp. 371–418

23. R.C. Merton, Option pricing when underlying stock returns are discontinuous. J. Financ. Econ. **3**(1–2), 125–144 (1976)
24. C.A. Ramezani, Y. Zeng, Maximum likelihood estimation of the double exponential jump-diffusion process. Ann. Financ. **3**(4), 487–507 (2007)
25. L.C.G. Rogers, Arbitrage with fractional Brownian motion. Math. Financ. **7**(1), 95–105 (1997)
26. G. Samoradnitsky, M.S. Taqqu, *Stable Non-Gaussian Random Processes: Stochastic Models with Infinite Variance*, vol. 1 (CRC Press, Boca Raton, 1994)
27. B.V. Shabat, *Introduction to Complex Analysis: Functions of Several Variables*, vol. 110 (American Mathematical Society, Providence, RI, 1992)
28. S.E. Shreve, *Stochastic Calculus for Finance II: Continuous-Time Models*, vol. 11 (Springer Science & Business Media, Berlin, 2004)
29. P. Tankov, *Financial Modelling with Jump Processes*, vol. 2 (CRC Press, Boca Raton, 2003)
30. J. Yeh, *Lectures on Real Analysis* (World Scientific, Singapore, 2000)

Chapter 6
Numerical Implementation and Parameter Estimation Under KOU Model

In this chapter, an overview of the different estimation methods, and the calculation of the parameters of the Kou model based on these estimations are presented. Among the used estimation methods are Maximum-Likelihood Estimation, Empirical Characteristic Function method, Generalized Method of moments, and Cumulant Matching method.

6.1 Estimation Method: Theoretical Background

In the following subsections, we will explain some estimation methods that have been used in this thesis.

6.1.1 Maximum-Likelihood Estimation

Maximum-Likelihood Estimation (MLE) can be expressed as the likelihood function of the given data. In other words the likelihood of a set of data is the probability of achieving that particular set of data within the given probability distribution model. Unknown model parameters are also incorporated in this expression. The values of these parameters that maximize the sample likelihood are known as the Maximum-Likelihood Estimates [15]. MLE method can be used in a large variety of statistical, data mining and optimization situations which makes it a consistent approach to parameter estimation problems. The choice of starting values affect the estimation, and the optimality characteristics may not be appropriate for small sample sets [15]. In finance, alternatives to the MLE approach have been used by practitioners. Despite its generality and well-known asymptotic properties, such as consistency, normality and efficiency, the likelihood function may not be tractable

A. Karimov, *Identifying Stock Market Bubbles*, Contributions to Management Science, https://doi.org/10.1007/978-3-319-65009-8_6

in many situations, namely, due to its boundlessness over the parametric space, instabilities or the existence of many local maxima [16]. Another obstacle with the MLE approach is that a closed form density cannot be attained for some families of distribution functions and, therefore, the MLE method will be computationally expensive when applied.

Let us denote by $S = (S_t)_{t\geq 0}$ the stochastic process of index values given at equally-spaced times: the one period rate of return $X_{\Delta t} = \ln S_t - \ln S_{t-\Delta t}$ assumed to be i.i.d., where Δt is the length of any subinterval of equally spaced observations. Let $\mathbf{X}$ denote the observed (random) return vector; then its PDF is:

$$f(\mathbf{X}, \boldsymbol{\theta}), \tag{6.1}$$

with $\boldsymbol{\theta} = (\theta_1, \theta_2, \ldots, \theta_m)^T$ being the vector of m unknown parameters that needs to be calculated.

The likelihood function can be calculated by the following product:

$$L(\mathbf{X}|\boldsymbol{\theta}) = \prod_{i=1}^{n} f(X_{i\Delta}, \boldsymbol{\theta}).$$

In writing the right hand side as the product of the density function assumptions have been made that the random sample variables are i.i.d. The log likelihood function is given by:

$$\ln L(\mathbf{X}|\boldsymbol{\theta}) = \sum_{i=1}^{n} \ln f(X_{i\Delta}, \boldsymbol{\theta}). \tag{6.2}$$

The MLE of $\boldsymbol{\theta}$ are obtained by maximizing the likelihood function. Since the maxima of this function are the same with that of the log-likelihood, and due to the monotonicity in $\boldsymbol{\theta}$ of natural logarithm function, we can maximize log-likelihood which is much simpler to work with than likelihood function.

6.1.2 Generalized Method of Moments

Generalized Method of Moments (GMM) is a generalization of the classical Method of Moments (MoM) estimation technique. MoM procedure equates population moments to sample moments in order to estimate population parameters. Since the introduction of GMM in 1982 by Hansen [10], it has been widely applied to analyze economic and financial data. Even though MLE is a more efficient estimator than GMM, the dependence of MLE on probability distribution can be a weakness. Some of these problems are sensitivity of statistical properties to the distributional assumption and computational burden [9]. In the GMM framework, the probability density function is not specified and this makes GMM a more computationally convenient method for parameter estimation. To employ the GMM to calculate parameters, the estimators are derived from so called moment conditions. A moment condition is a statement involving the data and the parameters [14]. For a dataset of X_t, where $t = 1, 2, \ldots, n$, drawn from a given probability distribution $\mathbb{P}$ and the parameter vector $\boldsymbol{\theta}_0$ satisfies the moment condition

$$\mathbb{E}\left[g(\mathbf{X}_t, \boldsymbol{\theta}_0)\right] = 0$$

for some known function g. In GMM, the basic idea is to construct the function g to form a valid moment condition, and the sample data are used to establish a sample analog of $\mathbb{E}\left[g(\cdot)\right]$ using the Law of Large Numbers [9]. A parameter $\hat{\boldsymbol{\theta}}$ is chosen to solve the next equation

$$\frac{1}{2}\sum_{t=1}^{n} \mathbb{E}\left[g(\mathbf{X}_t, \boldsymbol{\theta}_0)\right] = 0.$$

This allows us to consider the quadratic form

$$Q_t(\boldsymbol{\Theta}) = M_t(\boldsymbol{\theta})^T W_t M_t(\boldsymbol{\theta}), \tag{6.3}$$

where by $M_t(\boldsymbol{\theta})$ we indicate

$$M_t(\boldsymbol{\theta}) = \frac{1}{2}\sum_{t=1}^{n} \mathbb{E}\left[g(\mathbf{X}_t, \boldsymbol{\theta}_0)\right],$$

and W_t is a symmetric, positive semi-definite matrix which may depend on the data but it is required to converge in probability to a positive definite matrix for the estimator to be well defined. If $M_t(\boldsymbol{\theta})$ is a $q \times 1$ matrix, then W is a $q \times q$ matrix. The estimate $\hat{\boldsymbol{\theta}}$ is obtained by minimizing $Q_t(\boldsymbol{\theta})$. There are two main problems associated with the GMM, first one is which moments to match, and the second one is the number of moments included in estimation. Andersen [1] showed that the inclusion of an excessive number of moments produces bigger biases and larger root mean square errors. Hence, the employment of extra information may be harmful. We can conveniently derive all the moments via the characteristic function by taking advantage of the relationship between moments and cumulants. Denoting ϕ_x as a characteristic function of a random variable X and assuming that $\mathbb{E}|X|^n < \infty$, then ϕ_x has n continuous partial derivatives at $u = 0$, and we obtain for all $k = 1, 2, \ldots, n$, the subsequent representations,

$$M_k = \mathbb{E}\left[X_k\right] = \frac{1}{\mathrm{i}^k}\frac{\partial^k \phi_x(0)}{\partial u^k},$$

and

$$C_k = \frac{1}{\mathrm{i}^k} \frac{\partial^k \ln \phi_x(0)}{\partial u^k},$$

where M_k and C_k are the kth moment and kth cumulant, respectively.

6.1.3 Characteristic Function Estimation Method

Characteristic Function (CF) estimation method is applied in situations when the likelihood has more complicated form than the characteristic function because the characteristic function (CF) is always bounded and is available in a simpler form than the density in some important cases [16]. *Empirical Characteristic Function* (ECF) preserves all information in the data because there is an injective correspondence between the CF and cumulative distribution function due to the fact that CF is the Fourier—Stietjes transform of the CDF and this justifies the use of the ECF estimation method [21] and, therefore, an inference based on ECF can outperform the one which is based on generalized method of moments. Moreover, the final estimators will be consistent and asymptotically normal, under some conditions. In Lévy models, the CF is known via Lévy Khintchine theorem, given in below:

Theorem 6.1 (Lévy-Khinchin Representation [3]) *Let $(X_t)_{t\geq 0}$ be a Lévy process in $\mathbb{R}$ with characteristic triplet (A, ν, γ). Then, the characteristic function of X_t satisfies the relation*

$$\phi_{X_t}(u) = e^{t\psi(u)}, \quad u \in \mathbb{R}^d,$$

where $\psi(u)$ known as the characteristics exponent and given by

$$\psi(u) = i\gamma u - \frac{1}{2}Au^2 + \int_{\mathbb{R}^d} \left(e^{\mathrm{i}ux} - 1 - \mathrm{i}uxI_{|x|\leq 1}\right) \nu(dx),$$

where A is the diffusion component, $\gamma \in \mathbb{R}^d$ is the drift component and ν is a positive Radon measure on $\mathbb{R}^d \setminus \{0\}$ verifying:

$$\int_{|x|<1} |x|^2 \nu(dx) < \infty, \quad \int_{|x|\geq 1} \nu(dx) < \infty,$$

with ν being a Lévy measure of the jump distribution.

The ECF estimation method will be used to model log-returns as i.i.d. random variables using Kou's model. This method builds upon the works of Jiang and Knight [12] and Rockinger and Semenova [18] to evaluate the parameters of jump diffusion models. Former article used ECF method to calculate the parameters of affine jump diffusion models with latent variables while the second one employed the method to estimate the parameters of affine jump diffusion models with stochastic volatility. ECF method for i.i.d. random variables proposed by Heathcote [11] is used in this thesis.

6.1.3.1 Independent and Identical Distribution Case

Suppose that the PDF of X is defined as in Eq. (6.1), and $\boldsymbol{\theta} = (\theta_1, \theta_2, \ldots, \theta_m)^T$ are m unknown parameters that needs to be measured and denote by $\mathbf{X} = (X_1, X_2, \ldots, X_n)^T$ i.i.d. random variables, then the CF is specified by

$$\phi(u, \boldsymbol{\theta}) = \mathbb{E}[e^{iu\mathbf{X}}],$$

and the ECF is determined by

$$\hat{\phi}(u) = \frac{1}{n}\sum_{j=1}^{n} e^{iuX_j}.$$

By the Law of Large Numbers, $\phi(u)$ is a consistent estimator of $\hat{\phi}(u)$. The main idea used in ECF estimation is to minimize various distance measures between the ECF and CF. This method finds

$$\hat{\boldsymbol{\theta}} = \arg\min_{\boldsymbol{\theta}} \left\| \hat{\phi}(u) - \phi(u, \boldsymbol{\theta}) \right\|,$$

where $\|\cdot\|$ is usually a $\mathcal{L}^\infty$ or $\mathcal{L}^r$ weighted norm. In this thesis the $\mathcal{L}^2$ norm have been used. One can minimize

$$h(\boldsymbol{\theta}) = \int_{-\infty}^{\infty} \left| \hat{\phi}(u) - \phi(u, \boldsymbol{\theta}) \right|^2 g(u) du, \tag{6.4}$$

with $g(u)$ being a continuous weighting function. The choice of the weighting function $g(u)$ is often a concern. The inverse Fourier transform of the score function is the optimal weight function obtained by Feuerverger [7] using the Parseval identity, and it is given by

$$g(u) = \frac{1}{2\pi} \int_{-\infty}^{\infty} \exp(-iux) \frac{\partial \log f_{\boldsymbol{\theta}}(x)}{\partial \boldsymbol{\theta}},$$

depends on the density function. The resulting estimator attains maximum likelihood efficiency. We will define the exponential weighting function in the application part of this thesis (cf. Sect. 6.2).

6.1.3.2 Consistency and Asymptotic Normality

Consistency and asymptotic normality of the ECF estimators presented here follow from Heathcote [11]. The assumption here is that $h(\boldsymbol{\theta})$ can be differentiated under the integral sign.

$$\frac{\partial h(\boldsymbol{\theta})}{\partial \boldsymbol{\theta}} = \int_{-\infty}^{\infty} \frac{d}{d\boldsymbol{\theta}} \left|\hat{\phi}(u) - \phi(u,\boldsymbol{\theta})\right|^2 g(u)du. \tag{6.5}$$

The statistics $\hat{\boldsymbol{\theta}}$ minimizes

$$\begin{aligned} h(\boldsymbol{\theta}) &= \int_{-\infty}^{\infty} \left|\hat{\phi}(u) - \phi(u,\boldsymbol{\theta})\right|^2 g(u)du \\ &= \int_{-\infty}^{\infty} \left(\left[\operatorname{Re}\hat{\phi}(u) - \operatorname{Re}\phi(u,\boldsymbol{\theta})\right]^2 + \left[\operatorname{Im}\hat{\phi}(u) - \operatorname{Im}\phi(u,\boldsymbol{\theta})\right]^2\right) g(u)du. \end{aligned} \tag{6.6}$$

Here, $\hat{\phi}(u) = \operatorname{Re}\hat{\phi}(u) + \mathrm{i}\operatorname{Im}\hat{\phi}(u)$ and $\phi(u,\boldsymbol{\theta}) = \operatorname{Re}\phi(u,\boldsymbol{\theta}) + \mathrm{i}\operatorname{Im}\phi(u,\boldsymbol{\theta})$. The estimating equation becomes

$$\begin{aligned} \frac{\partial h(\boldsymbol{\theta})}{\partial \boldsymbol{\theta}} = -2\int_{-\infty}^{\infty} \Bigg(&\left[\operatorname{Re}\hat{\phi}(u) - \operatorname{Re}\phi(u,\boldsymbol{\theta})\right] \frac{\partial \operatorname{Re}\phi(u,\boldsymbol{\theta})}{\partial \theta_i} \\ &+ \left[\operatorname{Im}\hat{\phi}(u) - \operatorname{Im}\phi(u,\boldsymbol{\theta})\right] \frac{\partial \operatorname{Im}\phi(u,\boldsymbol{\theta})}{\partial \theta_i}\Bigg) g(u)du. \end{aligned} \tag{6.7}$$

Since $e^{\mathrm{i}uX_j} = \cos(uX_j) + \mathrm{i}\sin(uX_j)$, this implies that

$$\frac{1}{n}\sum_{j=1}^{n} e^{\mathrm{i}uX_j} = \frac{1}{n}\sum_{j=1}^{n} \left(\cos(uX_j) + \mathrm{i}\sin(uX_j)\right).$$

This means that $\operatorname{Re}\hat{\phi}(u) = \frac{1}{n}\sum_{j=1}^{n}\cos(uX_j)$ and $\operatorname{Im}\hat{\phi}(u) = \frac{1}{n}\sum_{j=1}^{n}\sin(uX_j)$. Equation (6.7) can be written as,

$$\begin{aligned} \frac{\partial h(\boldsymbol{\theta})}{\partial \boldsymbol{\theta}} = \frac{-2}{n}\int_{-\infty}^{\infty} \Bigg(&\left[\cos(uX_j) - \operatorname{Re}\phi(u,\boldsymbol{\theta})\right] \frac{\partial \operatorname{Re}\phi(u,\boldsymbol{\theta})}{\partial \theta_i} \\ &+ \left[\sin(uX_j) - \operatorname{Im}\phi(u,\boldsymbol{\theta})\right] \frac{\partial \operatorname{Im}\phi(u,\boldsymbol{\theta})}{\partial \theta_i}\Bigg) g(u)du. \end{aligned} \tag{6.8}$$

The estimator $\hat{\phi}(u)$ is the root of Eq. (6.8) for which $h'(\hat{\boldsymbol{\theta}}) > 0$.
The ECF estimator is consistent, i.e.,

$$\hat{\boldsymbol{\theta}} \xrightarrow{a.s.} \boldsymbol{\theta},$$

and asymptotically normally distributed,

$$\sqrt{n}(\hat{\boldsymbol{\theta}} - \boldsymbol{\theta}) \xrightarrow{d} N(\mathbf{0}, \mathbf{B}^{-1}(\boldsymbol{\theta})\mathbf{A}(\boldsymbol{\theta})\mathbf{B}^{-1}(\boldsymbol{\theta})), \quad n \to \infty, \tag{6.9}$$

where d in Eq. (6.9) stands for convergence in distribution, $\mathbf{A}(\boldsymbol{\theta})$ is the covariance matrix of the random variables

$$K^{(i)}(\boldsymbol{\theta}) = \int_{-\infty}^{\infty} \left(\left[\cos(uX_j) - \operatorname{Re}\phi(u,\boldsymbol{\theta})\right] \frac{\partial \operatorname{Re}\phi(u,\boldsymbol{\theta})}{\partial\theta_i} + \left[\sin(uX_j) - \operatorname{Im}\phi(u,\boldsymbol{\theta})\right] \frac{\partial \operatorname{Im}\phi(u,\boldsymbol{\theta})}{\partial\theta_i} \right) g(u)du,$$

for $i = 1, 2, \ldots, m$, given by

$$\mathbf{A}(\boldsymbol{\theta}) = \mathbb{E}\left[\frac{1}{n} \sum_{j=1}^{n} \sum_{h=1}^{n} K_j(\theta_i) K_h(\theta_i) \right].$$

Since X is a vector of i.i.d. observations, the above expression is given by:

$$\begin{aligned}
\mathbf{A}_{i,j}(\boldsymbol{\theta}) =& \frac{1}{n} \int_{-\infty}^{\infty} \int_{-\infty}^{\infty} \left[\frac{\partial \operatorname{Re}\phi(u,\boldsymbol{\theta})}{\partial\theta_i} \frac{\partial \operatorname{Re}\phi(s,\boldsymbol{\theta})}{\partial\theta_j} \operatorname{Cov}(\cos(u,X),\cos(s,X)) \right. \\
&+ 2 \frac{\partial \operatorname{Re}\phi(u,\boldsymbol{\theta})}{\partial\theta_i} \frac{\partial \operatorname{Im}\phi(s,\boldsymbol{\theta})}{\partial\theta_j} \operatorname{Cov}(\cos(u,X),\sin(s,X)) \\
&\left. + \frac{\partial \operatorname{Im}\phi(u,\boldsymbol{\theta})}{\partial\theta_i} \frac{\partial \operatorname{Im}\phi(s,\boldsymbol{\theta})}{\partial\theta_j} \operatorname{Cov}(\sin(u,X),\sin(s,X)) \right] e^{-u^2} e^{-s^2} duds,
\end{aligned}$$

where, from elementary trigonometric identities, for real numbers u, s we will get:

$$\operatorname{Cov}(\cos(u,X),\cos(s,X)) = \frac{1}{2}\left[\operatorname{Re}\phi(u-s,\boldsymbol{\theta}) + \operatorname{Re}\phi(u+s,\phi) - 2\operatorname{Re}\phi(u,\boldsymbol{\theta})\operatorname{Re}\phi(s,\boldsymbol{\theta})\right],$$

$$\operatorname{Cov}(\cos(u,X),\sin(s,X)) = \frac{1}{2}\left[\operatorname{Im}\phi(u+s,\boldsymbol{\theta}) - \operatorname{Im}\phi(u-s,\phi) - 2\operatorname{Re}\phi(u,\boldsymbol{\theta})\operatorname{Im}\phi(s,\boldsymbol{\theta})\right],$$

$$\operatorname{Cov}(\sin(u,X),\sin(s,X)) = \frac{1}{2}\left[\operatorname{Re}\phi(u-s,\boldsymbol{\theta}) - \operatorname{Re}\phi(u+s,\phi) - 2\operatorname{Im}\phi(u,\boldsymbol{\theta})\operatorname{Im}\phi(s,\boldsymbol{\theta})\right],$$

where the ith element in the vector $\boldsymbol{\theta} = (\theta_1, \theta_2, \ldots, \theta_m)^T$ is denoted by θ_i and $B(\boldsymbol{\theta})$ is the $m \times m$ symmetric matrix whose (i,j)-th entry is given by,

$$\mathbf{B}_{i,j}(\boldsymbol{\theta}) = \int_{-\infty}^{\infty} \left[\frac{\partial \operatorname{Re}\phi(u,\boldsymbol{\theta})}{\partial\theta_i} \frac{\partial \operatorname{Re}\phi(u,\boldsymbol{\theta})}{\partial\theta_j} + \frac{\partial \operatorname{Im}\phi(u,\boldsymbol{\theta})}{\partial\theta_i} \frac{\partial \operatorname{Im}\phi(u,\boldsymbol{\theta})}{\partial\theta_j} \right] e^{-u^2} du.$$

The method applied for approximating the integrals above is adaptive Gauss-Kronrod quadrature using MATLAB *quadgk* built-in function which attempts to approximate the integral of a scalar-valued function from a to b employing high-order global adaptive quadrature. The integral limits a and b can be $-\infty$ or ∞.

6.1.4 Monte-Carlo Simulation

A method for iteratively assessing a deterministic model by using as input the sets of random numbers is called Monte-Carlo simulation. This method is employed when the model is complicated, nonlinear, or comprising more than a couple uncertain parameters. Monte-Carlo methods are based on the analogy between probability and volume, and simulation can normally contain over 10,000 assessments of the considered model [8].

6.1.4.1 Principle of Monte-Carlo Simulation

The basic concept of Monte-Carlo simulation is as follows, assume that we are analyzing a random variable Y on a probability space $(\Omega, \mathcal{F}, \mathbb{P})$, which also notes the results of a conducted experiment. The repetitions of the experiment can be modeled by suggesting a sequence of random variables $Y_1, Y_2, \ldots, Y_n$, each having the same probability distribution as Y. Assuming that $Y_1, Y_2, \ldots, Y_n$ are independent, the sequence can be regarded as a model for repeated and independent runs for the experiment [2]. The Strong Law of Large Numbers shows that with certainty, we can derive the common expected values of the random variables.

6.1.4.2 Strong Law of Large Numbers

Theorem 6.2 ([6]) *Let $(Y_i)_{i \in \mathbb{N}}$ be a sequence of i.i.d. integrable random variables determined on the same probability space, such that for $i \in \mathbb{N}$, let $y = \mathbb{E}[Y_i]$, then*

$$\mathbb{P}\left(\lim_{n\to\infty} \frac{Y_1 + Y_2 + \ldots + Y_n}{n} = y\right) = 1.$$

The Strong Law of Large Numbers says that for almost every sample point $\omega \in \Omega$,

$$\frac{Y_1(\omega) + Y_2(\omega) + \ldots + Y_n(\omega)}{n} \to y \quad \text{as} \quad n \to \infty.$$

Therefore, if $Y_1, \ldots, Y_n$ is a sequence of random variables each of which has the same probability information as Y and $\mathbb{E}[Y] < \infty$, then

$$\frac{1}{n}\sum_{i=1}^{n} Y_i \overset{a.s.}{\to} \mathbb{E}(Y) \quad \text{as} \quad n \to \infty.$$

Monte-Carlo simulation has an advantage of being flexible compared to other numerical methods. Moreover, it serves as the only method of simulation in higher dimensions.

6.2 Estimation Methods: Numerical Application

6.2.1 *Characteristic Function and Moments of Kou Model*

Jump size in the Kou model has the double exponential distribution with the density given by Eq. (5.10) and the Lévy density being defined [3] as

$$\nu(y) = \lambda f_Y(y) = \lambda p\eta_1 e^{-\eta_1 y} 1_{y\geq 0} + \lambda q\eta_2 e^{\eta_2 y} 1_{y<0},$$

with Lévy triplet given by $\left(\mu, \sigma^2, \lambda f_Y(y)\right)$.

We have already introduced the Lévy-Khinchin representation in Theorem 6.1. For the case where the related process has a limited activity, i.e., in a limited period of time it has a finite number of jumps, then

$$\psi(z) = \mathrm{i}\gamma z - \frac{1}{2}Az^2 + \int_{-\infty}^{+\infty} (e^{\mathrm{i}zx} - 1)\nu(dx).$$

Applying Theorem 6.1 to the Kou model we get,

$$\begin{aligned}
\psi(z) =& \mathrm{i}\mu z - \frac{1}{2}\sigma^2 z^2 + \int_{-\infty}^{+\infty} (e^{\mathrm{i}zy} - 1)\nu(dy) \\
=& \mathrm{i}\mu z - \frac{1}{2}\sigma^2 z^2 + \int_{0}^{+\infty} (e^{\mathrm{i}zy} - 1)\lambda p\eta_1 e^{-\eta_1 y} dy + \int_{-\infty}^{0} (e^{\mathrm{i}zy} - 1)\lambda q\eta_2 e^{\eta_2 y} dy \\
=& \mathrm{i}\mu z - \frac{1}{2}\sigma^2 z^2 + \lambda \left(\frac{p\eta_1}{\eta_1 - \mathrm{i}z} + \frac{q\eta_2}{\eta_2 + \mathrm{i}z} - 1 \right).
\end{aligned}$$

Thus, the following holds

$$\phi_{X_t}(z) = \exp\left(t\left(\mathrm{i}\mu z - \frac{1}{2}\sigma^2 z^2 + \lambda \left(\frac{p\eta_1}{\eta_1 - \mathrm{i}z} + \frac{q\eta_2}{\eta_2 + \mathrm{i}z} - 1 \right)\right)\right). \tag{6.10}$$

The log-characteristic function $\Psi(z) = \ln\phi(z)$ which $\Psi(0) = 0$ is called a Cumulant Generating Function. To derive the cumulants we use the characteristic exponent calculated above, with the nth cumulants being defined by

$$C_n = \frac{1}{\mathrm{i}^n}\frac{\partial^n \Psi(0)}{\partial z^n}. \tag{6.11}$$

Applying the formula given in Eq. (6.11) to log-characteristic function gives the following population cumulants for Kou model:

$$\begin{aligned}
C_1 &= \Delta t\left(\mu - \frac{\sigma^2}{2} + \lambda\left(\frac{p}{\eta_1} - \frac{1-p}{\eta_2}\right)\right),\\
C_2 &= \Delta t\sigma^2 + 2\Delta t\lambda\left(\frac{p}{\eta_1^2} + \frac{1-p}{\eta_2^2}\right),\\
C_3 &= 6\Delta t\lambda\left(\frac{p}{\eta_1^3} - \frac{1-p}{\eta_2^3}\right),\\
C_4 &= 24\Delta t\lambda\left(\frac{p}{\eta_1^4} + \frac{1-p}{\eta_2^4}\right),\\
C_5 &= 120\Delta t\lambda\left(\frac{p}{\eta_1^5} - \frac{1-p}{\eta_2^5}\right),\\
C_6 &= 720\Delta t\lambda\left(\frac{p}{\eta_1^6} + \frac{1-p}{\eta_2^6}\right).
\end{aligned}$$

6.2.2 *Simulation of Kou Model*

We simulate at fixed set of dates $0 = t_0 < t_1 < \ldots < t_n$ without explicitly distinguishing the effects of the jump and diffusion terms, as specified by Glasserman [8]. Let us set $X_t = \ln S_t$ $(t = \{t_0, t_1, \ldots, t_n\})$, then the algorithm for the steps in a sequential Monte-Carlo procedure for Kou model will be as follows:

1. Simulate $Z \sim \mathcal{N}(0, 1)$,
2. Simulate $N \sim \mathcal{P}(\lambda\Delta t)$, where $\mathcal{P}$ stands for the Poisson distribution,
3. If $N \neq 0$, then simulate $\ln Y_1, \ldots, \ln Y_N$ and set $Jump = \ln Y_1 + \ldots + \ln Y_N$ else if $N = 0$, set $Jump = 0$. Since an exponential distribution is basically a Gamma distribution with shape and scale parameters of 1 and β, then $\ln Y_1 + \ldots + \ln Y_N$ will have a Gamma distribution with shape parameter N and scale parameter β, and the sign of $\ln Y_j$ is positive with probability p and negative with probability $1 - p$. In case if the Poisson random variable N takes the value n, the number of $\ln Y_j$ with positive sign obeys a Binomial distribution with parameters n and p.

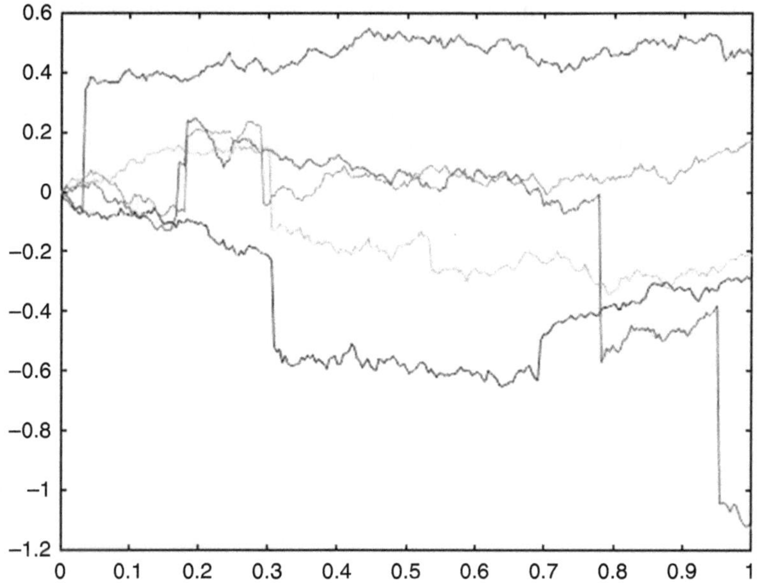

Fig. 6.1 Sample paths of Kou Model

Herewith,

(a) Generate $B \sim Binomial(N, p)$,
(b) Generate $G_1 \sim Gamma(B, \beta)$ and $G_2 \sim Gamma((N - B), \beta)$, and set $Jump = G_1 - G_2$,

4. Set

$$\begin{aligned} X(t_i + 1) =& X(t_i) + (\mu - \frac{1}{2}\sigma^2)\Delta t + \sigma\sqrt{\Delta Z + Jump}, \\ X_{\Delta t} =& (\mu - \frac{1}{2}\sigma^2)\Delta t + \sigma\sqrt{\Delta Z + Jump}. \end{aligned} \tag{6.12}$$

In $3(b)$, interpret a Gamma random variable with shape parameter zero as the constant 0 in case $B = 0$ or $B = N$. Some sample paths for simulated Kou model are shown in Fig. 6.1, the parameters chosen in the simulation are $\mu = 0$, $\sigma = 0.2$, $\eta_1 = 0.2$, $\eta_2 = 0.3$, $p = 0.5$, and $\lambda = 4.25$.

6.2.3 Cumulant Matching Method

The method of moment can be applied to evaluate the parameters of the Jump models, and these parameters are used as the point of origin in the subsequent methods of calculations. The procedure utilized in this thesis is an alternative version of the Method of Moments and it is named as *Cumulant Matching*. The cumulant catching method is based on the relationship between the population cumulant and the distribution parameters. Population cumulants are not known, the sample cumulants are used to estimate the parameters. Parameter estimation by cumulant matching is known to yield consistent estimators, but these estimators are not always efficient [17]. The cumulants are matched with the sample central moments because of the relationship that exists between them. The log-characteristic function $\psi(u) = \ln(\phi(u))$ is applied in generating the population cumulants C_k [3], with $k > 0$. Using this relationship between the central moment M_k and C_k, the first six sample cumulants of the models can be computed from the sample moments in the following way [13]:

$$\begin{aligned}
\hat{C}_1 &= M_1, \\
\hat{C}_2 &= M_2, \\
\hat{C}_3 &= M_3, \\
\hat{C}_4 &= M_4 - 3M_2^2, \\
\hat{C}_5 &= M_5 - 10M_3M_2, \\
\hat{C}_6 &= M_6 - 15M_4M_2 - 10M_3^2 + 30M_2^3,
\end{aligned}$$

where M_1 is the mean and M_2 being the second central moment is equal to the variance of the sample.

For Kou model, equating the population cumulants of the models to the sample cumulants gives the parameter estimates. As stated above, the parameters estimated by this method are consistent, but not always efficient. However, these estimates provide a good initial parameter for the algorithms of GMM and ECF. Nevertheless, the cumulant estimates may not exist or may have the wrong sign. In this thesis, the method proposed involves setting the population cumulants $C_k = \hat{C}_k$ as the sample cumulants, for $k = 1, \ldots, 6$. Solving these sets of equations without constraints might lead to getting values outside of the range that is desired; for instance, one could get a negative value for σ or a value greater than 1 or less than 0 for p. Therefore, we set

$$f_k := C_k - \hat{C}_k, \tag{6.13}$$

and the Sum of Squared Error, by $SSE := f_1^2 + f_2^2 + \ldots + f_6^2$; furthermore, the constrained optimization function *fmincon* from MATLAB is used to minimize SSE subject to the constraints that $\sigma > 0$, $\eta_1 > 1$, $\eta_2 > 0$ and $0 \leq p \leq 1$. SSE is at its minimum when $C_k \approx \hat{C}_k$. In particular, Kou model is achieved as described in the

Table 6.1 Parameters calculated using the Methods of Moments

μ	σ	η_1	η_2	p	λ
0.0902	0.3848	165.34395	2.1447	0.9101	86.0.2890

set of equation given in below:

$$
\begin{aligned}
f_1 =& \Delta t\left(\mu - \frac{\sigma^2}{2} + \lambda\left(\frac{p}{\eta_1} - \frac{1-p}{\eta_2}\right)\right) - M_1,\\
f_2 =& \Delta t\sigma^2 + 2\Delta t\lambda\left(\frac{p}{\eta_1^2} + \frac{1-p}{\eta_2^2}\right) - M_2,\\
f_3 =& 6\Delta t\lambda\left(\frac{p}{\eta_1^3} - \frac{1-p}{\eta_2^3}\right) - M_3,\\
f_4 =& 24\Delta t\lambda\left(\frac{p}{\eta_1^4} + \frac{1-p}{\eta_2^4}\right) - (M_4 - 3 - M_2^2),\\
f_5 =& 120\Delta t\lambda\left(\frac{p}{\eta_1^5} - \frac{1-p}{\eta_2^5}\right) - (M_5 - 10M_3M_2),\\
f_6 =& 720\Delta t\lambda\left(\frac{p}{\eta_1^6} + \frac{1-p}{\eta_2^6}\right) - (M_6 - 15M_4M_2 - 10M_3^2 + 30M_2^3).
\end{aligned}
$$

In Table 6.1, we provide the results of the optimization problem given as sum of squared errors and solved by *fmincon* function of MATLAB:

6.2.4 Maximum-Likelihood Estimation

The closed form of the transition density of Kou model is not known. Hence, the density is approximated using inverse Fourier transform of the characteristic function shown in Eq. (6.10):

$$
f_{X_{\Delta t}}(X) = \frac{1}{2\pi}\int_{-\infty}^{+\infty}(e^{-iuX}\phi_{X_{\Delta t}}(u))du = \frac{1}{\pi}\int_{0}^{+\infty}(e^{-iuX}\phi_{X_{\Delta t}}(u))du. \tag{6.14}
$$

After approximation of the right-hand side of Eq. (6.14) by using MATLAB *fmincon*, and getting estimation of parameters from CMM as initial parameter, we will minimize the negative value of the log-likelihood specified in Eq. (6.2), which is equivalent to maximizing the log-likelihood function. The parameters that maximize the log-likelihood function will also maximize the likelihood function and will be the MLE estimates for the proposed Kou model.

The integral given in Eq. (6.14) is evaluated using Matlab built-in function *quadgk*. In this thesis, the MATLAB built-in function *mle* is employed to evaluate

Table 6.2 Parameters calculated using the Maximum-Likelihood Estimation method

μ	σ	η_1	η_2	p	$1/\lambda$
1e−06	0.01649	5.00179	0.81393	0.68753	0.01898
(0.00064)	(0.00058)	(2.3055)	(0.41747)	(0.14257)	(0.00737)

the parameters of Kou model and *mlecov* is used to get the covariance matrix, applying the estimation of parameters, and the square roots of the diagonal entries in the covariance matrix give the standard error of the estimation. Parameter estimates for the model are shown in Table 6.2. The numbers in parenthesis are the standard deviations of the corresponding parameters.

6.2.5 Method of Characteristic Function Estimation

For Kou model, the characteristic function (CF) $\phi(\boldsymbol{\theta}, u)$ is defined in Eq. (6.10). The real part of the characteristic function is:

$$\begin{aligned}\mathrm{Re}(\phi) = &\exp\left[\delta t\left(-\frac{1}{2}\sigma^2 u^2 + \lambda\left(\frac{p\eta_1^2}{\eta_1^2+u^2} + \frac{q\eta_2^2}{\eta_2^2+u^2} - 1\right)\right)\right] \\ &\times \cos\left[\Delta t\left(u\mu - \frac{1}{2}u\sigma^2 + \lambda\left(\frac{p\eta_1^2}{\eta_1^2+u^2} + \frac{q\eta_2^2}{\eta_2^2+u^2}\right)\right)\right].\end{aligned}$$

The Imaginary part is represented by:

$$\begin{aligned}\mathrm{Im}(\phi) = &\exp\left[\delta t\left(-\frac{1}{2}\sigma^2 u^2 + \lambda\left(\frac{p\eta_1^2}{\eta_1^2+u^2} + \frac{q\eta_2^2}{\eta_2^2+u^2} - 1\right)\right)\right] \\ &\times \sin\left[\Delta t\left(u\mu - \frac{1}{2}u\sigma^2 + \lambda\left(\frac{p\eta_1^2}{\eta_1^2+u^2} + \frac{q\eta_2^2}{\eta_2^2+u^2}\right)\right)\right].\end{aligned}$$

In order to find the parameter vector $\boldsymbol{\theta} = (\mu, \sigma, \eta_1, \eta_2, p, \lambda)^T$, we will minimize the term in Eq. (6.4), given by Eq. (6.15) below, using parameters from MLE as initial parameters.

$$\int_{-K}^{K} |\varphi_{\boldsymbol{\theta}}(z) - \hat{\varphi}(z)|^2 w(z) dz, \tag{6.15}$$

where

$$\hat{\varphi}(z) = \frac{1}{t}\log\frac{1}{N}\sum_{k=1}^{N} e^{izX_k}$$

is the so-called Empirical Characteristic Exponent. Furthermore,

$$\varphi_{\boldsymbol{\theta}}(z) = \mathrm{i}\mu z - \frac{1}{2}\sigma^2 z^2 + \frac{p\eta_1\lambda}{\eta_1 - \mathrm{i}z} + \frac{q\eta_2\lambda}{\eta_2 + \mathrm{i}z} - \lambda$$

is the CF of the Kou model, and $w(z)$ is a weight function. Here, the set $(X_k)_{k=1,2,\ldots,N}$ is the dataset of log-returns used for calibration, and t is the period of these returns (e.g., $\sim \frac{1}{252}$ for daily returns using the unit 1 year, etc.).

Ideally, the weight function $w(z)$ should answer to the precision of $\hat{\varphi}(z)$ as an estimate of $\varphi_{\boldsymbol{\theta}}(z)$ for every z. Therefore it must be chosen as the reciprocal of the variance of $\hat{\varphi}(z)$:

$$\begin{aligned} w(z) &= \frac{1}{\mathbb{E}[(\hat{\varphi}(z) - \mathbb{E}[\hat{\varphi}(z)])\overline{(\hat{\varphi}(z) - \mathbb{E}[\hat{\varphi}(z)])}} \\ &= \frac{t^2 \varphi_{\theta^*}(z)\overline{\varphi_{\boldsymbol{\theta}^*}(z)}}{\mathbb{E}[(\hat{\varphi}(z) - \varphi_{\theta^*}(z))\overline{(\hat{\varphi}(z) - \varphi_{\boldsymbol{\theta}^*}(z))}}, \end{aligned}$$

with $\boldsymbol{\theta}^*$ being a true parameter. Nevertheless, this definition is not fruitful to deal with, as it relies upon the unknown parameter vector $\boldsymbol{\theta}$ and therefore cannot be calculated. Since the return distribution is somewhat close to Gaussian, the CF of the weight w may be calculated with a Gaussian one,

$$w(z) \approx \frac{e^{-\sigma_*^2 z^2}}{1 - e^{-\sigma_*^2 z^2}},$$

where $\sigma_*^2 = Var(X_k)$, $k = 1, 2, \ldots, N$, is the log returns data variance. The cut-off parameter K in Eq. (6.15) has be chosen to be equal to 60, based on tests with simulated data [4].

Obtained parameter estimations of the Kou model are displayed in Table 6.3. The values for the parameters depend on the estimation method used. For Kou Model, estimation based on ECF saves time, because of Fourier inversion involved in using the MLE estimation method.

Figure 6.2 illustrates the goodness of our fit achieved. Solid line in the figure indicates the Kernel density estimator applied directly on the data. Dashed line on the other hand is the Kernel density estimator derived from the Kou model simulation with parameters calculated via ECF. We note that indeed Kou model fits the smoothed returns density very well.

Table 6.3 Parameters calculated using the Empirical Characteristic Function method

μ	σ	η_1	η_2	p	λ
−0.34970	0.15743	60.67018	53.45101	0.65765	86.19877

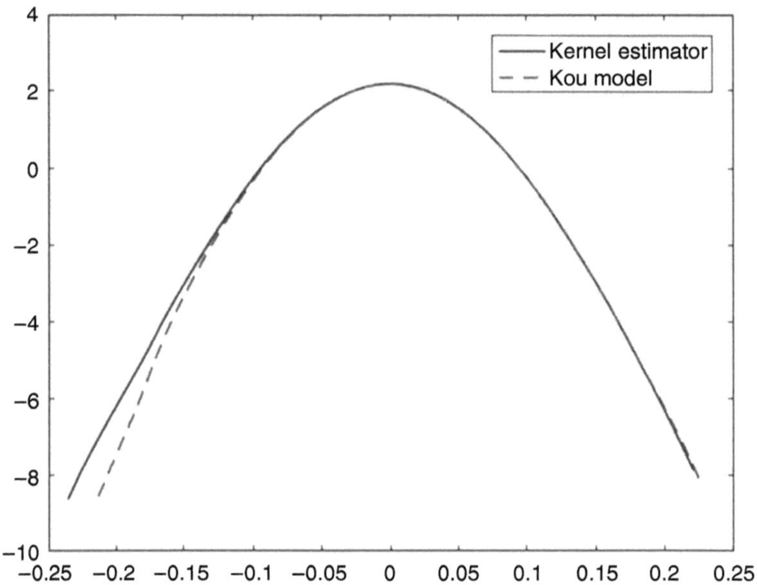

Fig. 6.2 Logarithm of the density for *S&P 500* Index time series

For the sake of assessing the goodness of fit of the Kou distributions to the market index data from another point of view, we use the quantile-quantile (Q-Q)-plot. A Q-Q plot is a plot of the quantiles of two distributions against each other. The order of the points in the plot is chosen to analyze the two distributions. If the plotted points lie roughly on the line $y = x$, then the compared distribution fits the data well. In order to show the goodness of fit using Q-Q plot, the parameters obtained by ECF from logreturns are used to simulate the distributions of the models, and then the quantiles of the distributions are compared to the quantiles of historical log-return data sets. Q-Q plots for our models are provided in Fig. 6.3. For the model based on the normal distribution, the deviation from the straight line is clearly seen (cf. the right panel). The Q-Q plots of the simulated returns against the historical returns show that Kou models do significantly better when compared to the models based on normal distribution. The quantiles of the simulated distributions are much more aligned with the quantiles of the historical return distribution than it was the case for the plain normal distribution.

One-sample and Two-sample Kolmogorov-Smirnov (KS) statistic goodness-of-fit tests are also employed to elaborate the null hypothesis that the log-returns have the normal distribution, and to test whether the empirical distribution F_{emp} and the fitted distribution F_{fit} are sampled from the same distribution, respectively. MATLAB's one-sample Kolmogorov-Smirnov test, $[h, p, ksstat, cv] = kstest(x)$, is applied to compare the empirical data X to the standard normal distribution. The null hypothesis says that the distribution function of X is the same as the standard normal distribution. The alternative hypothesis is that the distribution function of X

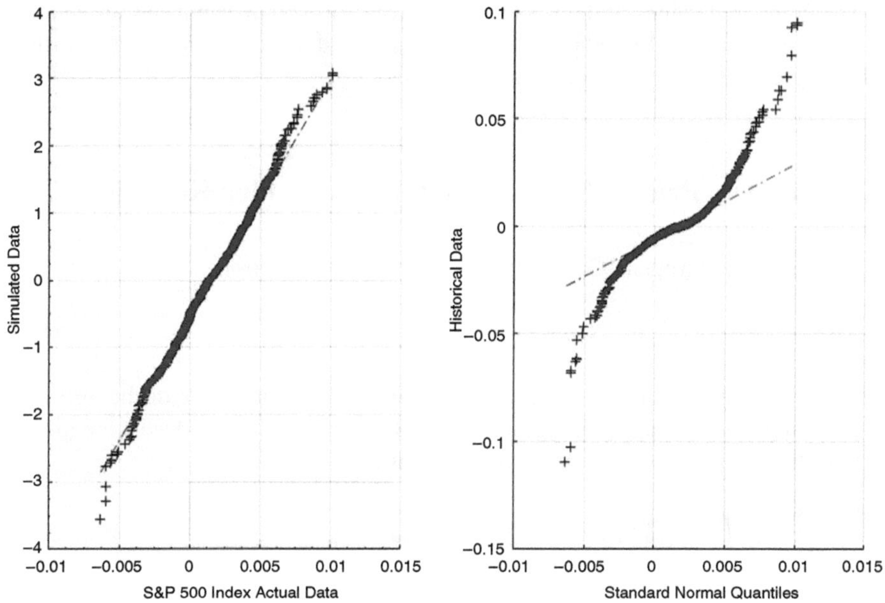

Fig. 6.3 The Q-Q plot of Kou fitted *S&P 500* Index Data vs. Historical Data (*left panel*) and Q-Q plot of Historical Data vs. Normal Density (*right panel*)

Table 6.4 Kolmogorov-Smirnov distance and probability Kou distribution

	KS distance	Probability
One sample test	0.49675	8.19e−236
Two sample test	0.02071	0.94414

significantly different from the standard normal distribution. In case when the value of h is equal to 1 the null hypothesis can be rejected at the 5% significance level, and if value of h is equal to 0 then the test will fail to reject the null hypothesis. The null hypothesis is accepted if p is greater than 5% and rejected otherwise. As shown in Table 6.4, the null hypothesis is rejected. In order to contrast the distributions of the two data vectors we will employ the two-sample Kolmogorov-Smirnov test. The test uses the empirical data X_1 and the fitted data X_2 to check whether the empirical cumulative distribution function F_{emp} and the fitted cumulative distribution function F_{fit} are sampled from the same distribution. The null hypothesis consists of the statement: X_1 and X_2 are from the same continuous distribution. The alternative hypothesis is that these samples are from distant continuous distributions. As in the one-sample test the null hypothesis will be rejected the 5% significance level if h is 1, and the test will fail to reject the null hypothesis if h is 0. The test statistic is:

$$KS = \max_{y \in \mathbb{R}} \left| F_{emp}(y) - F_{fit}(y) \right| .$$

The compared values of the KS test at $\alpha = 5\%$ for Kou distributions are reported in Table 6.4, the null hypothesis is accepted if the distance is too large and the p value is greater than α, and rejected otherwise.

6.3 Bid-Ask Prices of European Options Under Kou Model

The price of the European Call Option is given by the function

$$C_T = \max\{(S_T - K), 0\} \tag{6.16}$$

with S being the index price, K the strike price, and T the maturity of the option.

By using Eq. (4.4) apply integration by parts formula we can derive the general form of the bid price of the European call option:

$$\begin{aligned} b_\gamma(C) &= \int_0^\infty x\, d\Psi^\gamma(F_{C_T}(x)) \\ &= \int_K^\infty (x-K)\, d\Psi^\gamma(F_{S_T}(x)) \\ &= (x-K)\,(\Psi^\gamma(F_{S_T}(x)) - 1)|_K^\infty + \int_K^\infty (1 - \Psi^\gamma(F_{S_T}(x)))\; d(x-K) \\ &= \int_K^\infty (1 - \Psi^\gamma(F_{S_T}(x)))\; dx. \end{aligned} \tag{6.17}$$

For the ask price of call option, we have to note that for $x > 0$:

$$\begin{aligned} F_{-C}(x) &= \mathbb{P}(-(S-K) \le x) = \mathbb{P}((S-K) \ge -x) \\ &= \mathbb{P}(S \ge K - x) = 1 - \mathbb{P}(S \le K - x) = 1 - F_S(K-x). \end{aligned} \tag{6.18}$$

Now by using Eq. (4.3) and applying integration by parts formula, we can derive the general form of the ask price of the European call option:

$$\begin{aligned} a_\gamma(C) &= -\int_{-\infty}^0 x\, d\Psi^\gamma(F_{-C_T}(x)) \\ &= -\int_{-\infty}^0 x\, d\Psi^\gamma(1 - F_{S_T}(K-x)) \\ &= -\int_0^\infty x\, d\Psi^\gamma(1 - F_{S_T}(K+x)) = -\int_K^\infty (x-K)\, d\Psi^\gamma(1 - F_{S_T}(x)) \end{aligned}$$

$$= -\Psi^\gamma(1 - F_{S_T}(x))(x-K)|_K^\infty + \int_K^\infty \Psi^\gamma(1 - F_{S_T}(x))\, d(x-K)$$

$$= \int_K^\infty \Psi^\gamma(1 - F_{S_T}(x))\, dx. \tag{6.19}$$

Analogically, we can derive the general form of the ask price of the European put option. It is well recognized that the price of the European put option at maturity is defined by the following formula:

$$P_T = \max\{(K - S_T), 0\}, \tag{6.20}$$

with S_T being the index price at maturity time, K the strike price, and T the maturity time of the option.

Again by using Eq. (4.4) and employing the integration by parts formula, we can derive the general form of the bid price of the European put option:

$$\begin{aligned} b_\gamma(P) &= \int_0^\infty x\, d\Psi^\gamma(F_{P_T}(x)) \\ &= \int_K^\infty x\, d\Psi^\gamma(1 - F_{S_T}(K-x)) = -\int_0^K (K-x)\, d\Psi^\gamma(1 - F_{S_T}(x)) \\ &= -(K-x)\,(\Psi^\gamma(1 - F_{S_T}(x)) - 1)|_0^K + \int_0^K (1 - \Psi^\gamma(1 - F_{S_T}(x)))\, d(K-x) \\ &= \int_0^K (1 - \Psi^\gamma(1 - F_{S_T}(x)))\, dx. \end{aligned} \tag{6.21}$$

For the ask price of put option, we have to note that for $x > 0$:

$$\begin{aligned} F_{-P}(x) &= \mathbb{P}(-(K-S) \le x) = \mathbb{P}((K-S) \ge -x) \\ &= \mathbb{P}(S \le K + x) = F_S(K+x). \end{aligned} \tag{6.22}$$

Now, by using Eq. (4.3) and applying integration by parts formula, we arrive at the general form of the ask price of the European put option:

$$\begin{aligned} a_\gamma(P) &= -\int_{-\infty}^0 x\, d\Psi^\gamma(F_{-P_T}(x)) \\ &= -\int_{-\infty}^0 x\, d\Psi^\gamma(F_{S_T}(K+x)) \\ &= -\int_0^\infty x\, d\Psi^\gamma(F_{S_T}(K-x)) = \int_0^K (K-x)\, d\Psi^\gamma(F_{S_T}(x)) \end{aligned} \tag{6.23}$$

Table 6.5 Bid and ask prices of European options under the double-exponential Kou model

Option	Price
b_γ (C)	$\int_K^\infty \left(1-\Psi^\gamma(F_{S_T}(x))\right)\,dx$
a_γ (C)	$\int_K^\infty \Psi^\gamma(1-F_{S_T}(x))\,dx$
b_γ (P)	$\int_0^K \left(1-\Psi^\gamma(1-F_{S_T}(x))\right)\,dx$
a_γ (P)	$\int_0^K \Psi^\gamma(F_{S_T}(x))\,dx$

$$= \Psi^\gamma(F_{S_T}(x))(K-x)|_0^K + \int_0^K \Psi^\gamma(F_{S_T}(x))\,d(x-K)$$

$$= \int_0^K \Psi^\gamma(F_{S_T}(x))\,dx.$$

The general formulas of the options are provided in Table 6.5.

We have already calculated straightforwardly the form of the distribution function of S_T, $F_{S_T}(x)$, and the distortion function that we are going to use in this calculation will be *minmaxvar* function as it is similar to the Wang distortion function that has been applied in the Brownian case. This distortion function possesses the following representation:

$$\Psi^\gamma(u) = 1 - \left(1 - u^{\frac{1}{1+\gamma}}\right)^{1+\gamma}, \quad u \in [0,1]; \ \gamma \geq 0. \tag{6.24}$$

6.4 Data and Numerical Application of the Estimation Results of Kou Model

In this part we will employ the formulas at Table 6.5 to real data and derive the daily values of the γ. As in Sect. 4.4, we use the data of European put and call options written on *S&P 500* index. Again the chose of this index option is made mainly because it gives a general overview of the North-American option market that is one of the effective option markets in the world. Additionally, due to the nature of the *S&P 500* index we may assume that the company-specific events will have minor effects only on the calculation of the values of γ, and all the effects will come from the impacts of main events in the market, such as the financial crisis.

Expressions of bid and ask prices of the European options in Table 6.5 are easy to estimate numerically, if the distribution function of S_T is given. The formulas related to the put option, given by integrals with bounded limits, can be further simplified by employing Simpson's Rule for the numerical approximation of the integral

$$\int_a^b f(x)dx \approx \frac{\Delta x}{3}\left[f(x_0) + 4f(x_1) + 2f(x_2) + \ldots + 2f(x_{n-2}) + 4f(x_{n-1}) + f(x_n)\right], \tag{6.25}$$

with $\Delta x = \frac{b-a}{n}$, n being the even number of the subintervals. For the formulas related with call options, firstly we will have to make a change of variables of $x = \frac{1}{t}$, and, afterward, we will employ the method of Gauss-Legendre approximation.

The values of γ on a dataset are again estimated by minimizing the total-squared error (TSE), where TSE is the sum of the squared variations between the market prices and the theoretical prices:

$$TSE_{bid}(\gamma) = \sum_{i=1}^{\tau} \left(bid_i - b_{\gamma,i}\right)^2, \tag{6.26}$$

$$TSE_{ask}(\gamma) = \sum_{i=1}^{\tau} \left(ask_i - a_{\gamma,i}\right)^2, \tag{6.27}$$

or

$$TSE_{bid,ask}(\gamma) = \sum_{i=1}^{\tau} \left(\left(bid_i - b_{\gamma,i}\right)^2 + \left(ask_i - a_{\gamma,i}\right)^2\right), \tag{6.28}$$

with τ being the number of days for which illiquidity premium will be calculated, for example for daily γ calculations $\tau = 1$, for weekly γ calculation $\tau = 5$, and so on. Minimizing TSE delivers the market level's γ. We use Eq. (6.28) and the minimization problem can be given as in below:

$$\begin{aligned} &\underset{\gamma}{\text{minimize}} \quad TSE_{bid,ask}(\gamma) \\ &\text{subject to} \quad \gamma \geq 0. \end{aligned} \tag{6.29}$$

The parameter γ can be estimated hourly, daily, weekly, monthly or yearly as a constant value. We have estimated daily γ values. One of the interesting characteristics of the dataset is that it covers the period of the latest financial crises. It is essential to know the effects of a crisis on the acceptability level of the derivative that needs to be priced. These effects can be taken into consideration by adjusting the expected level of γ.

Figure 6.4 reflects the daily values of the Illiquidity Premium, calculated using the Kou model.

Figure 6.5 depicts the differences between the illiquidity premium, calculated by using the Kou and Black-Scholes Models.

There are several noticeable differences between the illiquidity premiums calculated from different models. The first one is the smoothness of the illiquidity premium derived from Kou model. In our opinion this is the case due to the fact that the Black-Scholes model does not take into account the jumps which have been proved to be the part of the movements of financial instruments. Kou model, on the contrary, by employing a double-exponential jump process takes into account the jumps related to the movements stock market price.

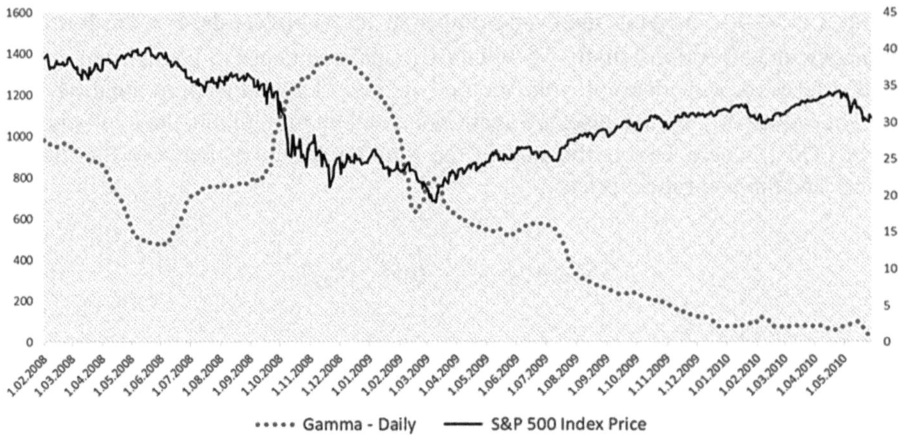

Fig. 6.4 *S&P 500* Index and Daily Illiquidity Premium

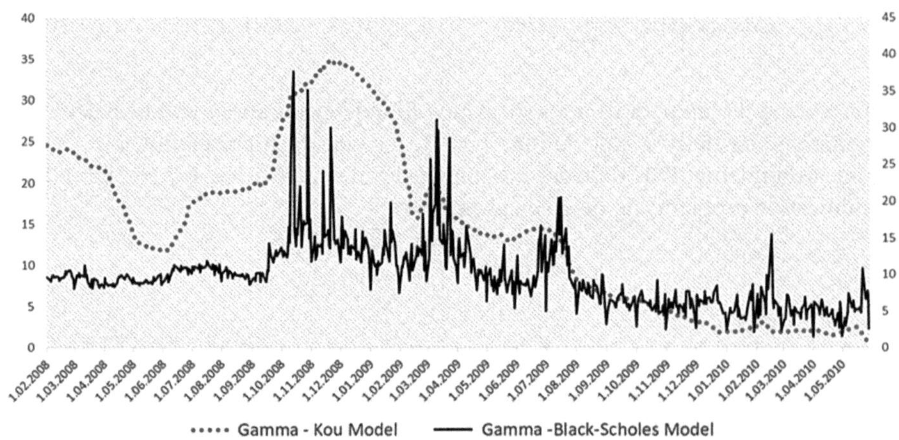

Fig. 6.5 Comparison of Illiquidity Premium Derived from Kou and Black-Scholes Models

Secondly, the premium calculated from the Kou model starts with high values; then it declines until the start of the Mortgage crisis, but the Black-Scholes model does not have this feature as the illiquidity premium has a nearly constant value initially. This is due to the fact that on 9th of August 2007, BNP Paribas, one of the Systematically Important Financial Institutes (SIFI), announced that it ceased activity of number of its hedge funds specialized over the mortgage debt. This was the moment when it became clear that investors and bankers were incorrect regarding the prices of more than trillions of dollars worth of derivative contracts, and they were worth a lot less than these investors had imagined [5]. This action of BNP Paribas began the seizure in the banking system and, moreover, in the overall economy. Around then no one knew how enormous the losses were or how big the exposure of individual banks to the crisis really was; thus, trust dissipated overnight and banks quit working with each other, which send market and economy to fluctuate.

Thirdly, after a couple of months of low values, between May and June 2008, which we think is the result of the calming of the market after the BNP Paribas news, the illiquidity premium started to rise, which occurred around the beginning of the third quarter of the year 2008. As we have mentioned, the premiums calculated from the Black-Scholes model are flat during the first 3 quarters and show an increase only towards the end of the third quarter of the 2008. The nearly one-quarter of difference is related with the bankruptcy of the Lehman Brothers. On 15th of September 2008 the investment bank Lehman Brothers went bankrupt, as the U.S. government did not bail it out. Before the bankruptcy, on 9th of June 2008, Lehman Brothers reported a second-quarter loss of $2.8 billion [20]; moreover the Lehman's quest to find a capital infusion created many speculations regarding them, and so the uncertainty in the market increased, hence the illiquidity premium went up.

At the point when Lehman Brothers went bankrupt, the thought that all banks were "too big to fail" no more remained true, and the outcome was that banks were considered to be dangerous and risky. Within a month, the danger of a domino effect through the worldwide monetary system constrained western governments to infuse immeasurable amounts of capital into their banking system to counteract them from collapsing. The banks were saved at the last possible second, however it was past the point where it is possible to keep the worldwide economy from going into freefall. This came after a period when high oil prices had influenced national banks that the need was to keep interest rates high as a defense against expansion instead of to cut them in expectation of the financial crisis spreading to the real economy [5].

From Fig. 6.5 we may also see that there are two more peaks for the results of the Kou model, namely, in early 2009 and in mid 2009. These peaks are also true for the Black-Scholes model results. The first peak in both models that has occurred in early 2009 is related with the General Motors (GM) and Chrysler solicitation of emergency loans with a specific end goal to address approaching money deficiencies following dramatic drops in automobile deals all through 2008. By April 2009, the circumstance had intensified to such an extent that both GM and Chrysler were confronted with approaching Chap. 11 and liquidation. With the aim to avoid unemployment increase and a destabilizing harm to the whole assembling segment, the U.S. also, Canadian governments gave extraordinary financial bailout ($85 billion) help in order for organizations to restructure [19]. Both organizations independently petitioned for this protection by June 1, 2009, and this is where uncertainty in the market started again, which corresponds to the second peak in Fig. 6.5.

References

1. T.G. Andersen, B.E. Sørensen, GMM estimation of a stochastic volatility model: a monte carlo study. J. Bus. Econ. Stat. **14**(3), 328–352 (1996)
2. M. Briani, Numerical methods for option pricing in jump-diffusion markets, Ph.D. thesis, Università degli Studi di Roma "La Sapienza", 2003
3. R. Cont, P. Tankov, *Financial Modelling with Jump Processes* (Chapman & Hall, Boca Raton, FL, 2004)
4. R. Cont, P. Tankov, Constant proportion portfolio insurance in the presence of jumps in asset prices. Math. Financ. **19**(3), 379–401 (2009)
5. L. Elliott, Global financial crisis: five key stages 2007–2011 (2011), https://www.theguardian.com/business/2011/aug/07/global-financial-crisis-key-stages
6. N. Etemadi, An elementary proof of the strong law of large numbers. Z. Wahrsch. verw. Gebiete **55**(1), 119–122 (1981)
7. A. Feuerverger, P. McDunnough, On some fourier methods for inference. J. Am. Stat. Assoc. **76**(374), 379–387 (1981)
8. P. Glasserman, *Monte Carlo Methods in Financial Engineering*, vol. 53 (Springer Science & Business Media, New York, 2003)
9. A.R. Hall, in *Generalized Method of Moments*. Advanced Texts in Econometrics Series (Oxford University Press, Oxford, 2005)
10. L.P. Hansen, Large sample properties of generalized method of moments estimators. Econometrica **50**, 1029–1054 (1982)
11. C. Heathcote, The integrated squared error estimation of parameters. Biometrika **64**(2), 255–264 (1977)
12. G.J. Jiang, J.L. Knight, Estimation of continuous-time processes via the empirical characteristic function. J. Bus. Econ. Stat. **20**, 198–212 (2012)
13. J. Kenney, E. Keeping, in *Cumulants and the Cumulant-Generating Function*. Mathematics of Statistics (Van Nostrand, Princeton, NJ, 1951)
14. H.B. Nielsen, *Generalized Method of Moments (GMM) Estimation*. Econometrics (2005)
15. N. Nist, *Sematech e-Handbook of Statistical Methods* (National Institute of Standards and Technology, Gaithersburg, MD, 2006)
16. P. Olivares, L. Seco, Stable distributions: a survey on simulation and calibration methodologies. Technical Report (2003). http://www.risklab.ca/ewExternalFiles/R5.pdf
17. S.J. Press, A compound events model for security prices. J. Bus. **40**, 317–335 (1967)
18. M. Rockinger, M. Semenova, Estimation of jump-diffusion processes via empirical characteristic functions (2005). Available at SSRN 770785
19. J. Van Biesebroeck, T.J. Sturgeon, Effects of the 2008–09 crisis on the automotive industry in developing countries: a global value chain perspective, in *Global Value Chains in a Postcrisis World* (The World Bank, Washington, DC, 2010), pp. 209–244
20. R.Z. Wiggins, T. Piontek, A. Metrick, The Lehman brothers bankruptcy a: overview, Yale Program on Financial Stability Case Study (2014)
21. J. Yu, Empirical characteristic function estimation and its applications. Econ. Rev. **23**(2), 93–123 (2004)

Chapter 7
Illiquidity Premium and Connection with Financial Bubbles

In this chapter, we present an early-warning signaling for financial bubbles by benefiting from the theory of Conic Finance, from optimization and suitably chosen numerical methods. We consider the U.S. markets since bubbles, mostly, occur in this market.

7.1 A Brief History of Financial Bubbles

Here, we present famous financial bubbles which occurred in the history, in order to perceive destructive results of a possible forthcoming bubble in present times and future.

7.1.1 Tulip Mania

The first well-known bubble in economic history is the tulip mania that originated in the Republic of Netherlands in the 1630s. The nation was encountered an extraordinary flourishing in the sixteenth and seventeenth century, during which the tulip was brought into Europe from the Ottoman Empire. The bloom quickly turned into a desired extravagance thing and a symbol of status as a result of its magnificence and the abundance of its varieties. Countless speculators began trading tulip and, therefore, the cost of the tulip bulbs raised to a mind blowing high levels. For instance, the knob of "Viceroy" costed around 3000 and 4200 florins; at the same time, a talented expert earned around 300 florins a year [25]. In February 1637, the tulip market suddenly broke down [39] abruptly, as shown in Fig. 7.1.

A. Karimov, *Identifying Stock Market Bubbles*, Contributions to Management Science, https://doi.org/10.1007/978-3-319-65009-8_7

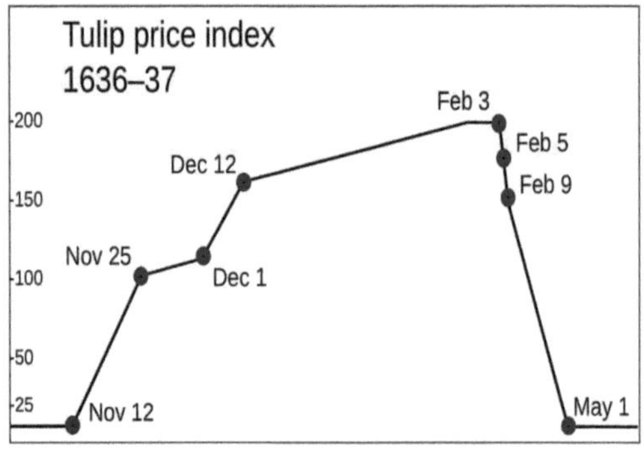

Fig. 7.1 Tulip prices between 1636 and 1637 [23]

7.1.2 South-Sea Bubble

The South-Sea bubble which originated in 1720 was an incredible financial bubble, caused by the stock speculation in the South-Sea Company. Along with the War of the Spanish Succession, a lot of the British government obligations were issued, and the administration needed to remove the interest rates of the obligation to alleviate its financial pressure. In the meantime, the stock of South-Sea organization was exceptionally well known in the light of the fact that it was given a trade monopoly with Spain's South American provinces as an element of a treaty throughout the War of Spanish Succession. The South-Sea Company wanted to hedge its risks by purchasing the government obligations with its overvalued stocks and get a steady income. Under this condition, the South-Sea scheme was put into action precisely the same way as our arguments above. This plan was thought to be a win-win trading. As an outcome, people in general began to purchase the stocks of South-Sea Company and the illicit activities from the organization (misrepresentation, loaning cash to the purchasers to empower the purchase of their stocks, etc.) increased the irrational behavior.

As Fig. 7.2 shows, the offer cost had ascended from the time when the plan was suggested: from 128 pounds in January, 1720, to 1000 pounds toward the beginning of August, 1720, trailed by a sensational tumble down to around 100 pounds during several months. Many investors lost a huge amounts of cash, along with Sir Isaac Newton. When he was given a question regarding the continuance of the rising of South-Sea stock, he replied: "I can calculate the movement of the stars, but not the madness of men" [38].

7.1.3 1929 Great Depression

The end of World War I introduced a new era to America. A time of certainty, good faith and welfare was experienced by the people of the United States in the 1920s. After World War I, industrialization and development of new technologies,

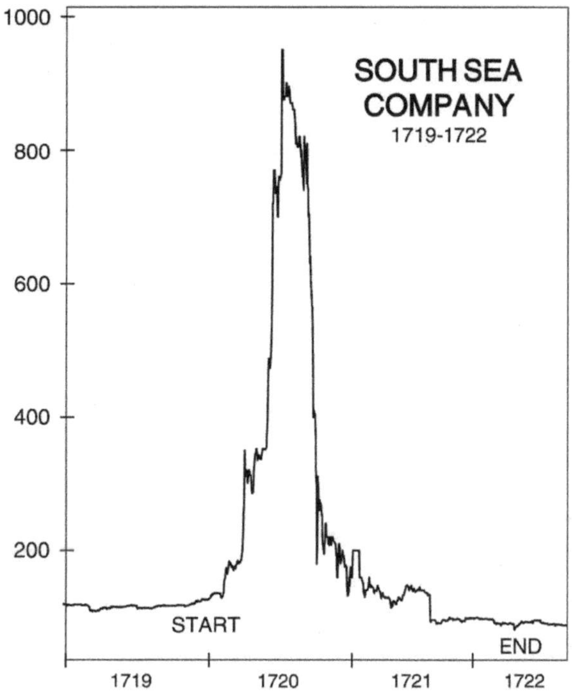

Fig. 7.2 South sea company share prices [23]

for example, of radio, automobile and air flight, supported the economic and social blast. The Dow Jones Industrial Average, DJIA, expanded all through the 1920s and due to the nation's solid economic conditions, the greater part of the financial experts believed that shares were the most certain investment. These brought about numerous investors to purchase stocks, greedily [8]. After a while, speculators acquired stocks on margin. In those years, just 10–20% of the stock cost were paid by the purchaser and, thus, 80–90% of the expense of the stock cost were being paid by the broker. Given the chance that the cost of the shares declined lower than the loan amount, the agent or broker would most likely issue a margin call, i.e., the purchaser had to pay back his loan as cash instantly. Along these lines, to purchase shares on margin could be exceptionally unsafe. Nonetheless, in the 1920s, numerous investors appeared who anticipated to make a profit on the stock market effectively, called speculators, in order to buy these stocks on margin. They assumed that this ascending process in prices would never end; so they could not perceive how genuine the danger was which they were getting in [31].

So, after all of these speculations, the Dow Jones Industrial Average had increased from 60 to 400 between 1921 and 1929. This produced a considerable number of new millionaires. Numerous individuals sold their homes and put their savings into securities at the exchange markets. In any case, few individuals truly knew about the companies in which they invested [8].

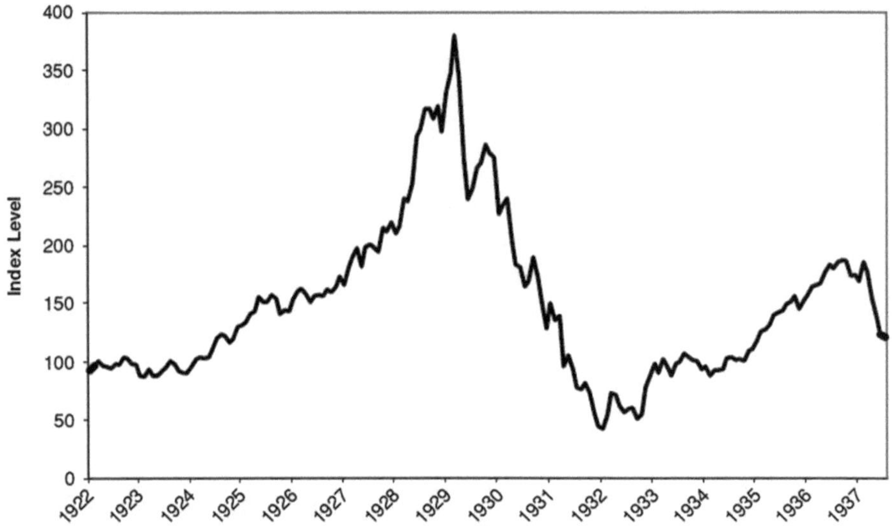

Fig. 7.3 Dow Jones Industrial Average between 1922 and 1937 (source: Thomson Reuters Eikon)

In 1929, from June through August, stocks prices saw some highest peaks. Economist Irving Fisher expressed that "Stock prices have achieved what resembles an all time high level"; this was just the comment numerous speculators needed to hear and believe. On 3rd of September 1929, the closing price of DJIA index was at 381.17 and after 2 days, the market started falling. Stock prices vacillated all through September and into October [31].

As demonstrated in Fig. 7.3, a steady bear market, a market in which the prices of shares were diminishing, had begun by October 1929. At 24th of October 1929 which is known as Black Thursday, alarm offering began since investors recognized that the stock market boom was indeed an over-inflated speculative bubble [8]. Despite the fact that the Federal Reserve Bank raised interest rates a few times to alleviate the uncertainty in securities' exchange and the overheated economy in 1929, this could not prevent from a tragic end. At the point when stock markets crashed at 28th and 29th of October, millionaire margin investors went bankrupt right away. In November 1929, DJIA sharply declined from 400 to 145. Over $5 billion worth of business sector capitalization had vanished from stocks that were traded at the New York Stock Exchange in only 3 days. The market crash of 1929 brought an extraordinary economic crisis, known as the Great Depression [8].

7.1.4 The Tech Bubble

By the mid 1990s, PCs were turning out to be progressively common for both business and individual use. PCs already become to be truly helpful business instruments that conceded their users a huge support in productivity. Business applications were developed to help end users with a diversity of assignments from bookkeeping via expense planning to word processing.

In the course of the 1990s, the U.S. PC industry chose to concentrate fundamentally upon software as opposed to designing and assembling PC equipment. The purpose behind this focus was founded in the reason that PC software was an item with high overall revenues, dissimilar to PC equipment. Software organizations' stocks were extremely strong performers all through the 1990s. Enthusiasm over the software business prompted the making of numerous small software startups, with a decent part of these organizations being pushed by undergrads in garages. For all intents and purposes each software startup wanted to become "the Next Microsoft".

By the mid 1990s, the index of technology stocks, NASDAQ, was ascending at a high pace, making numerous tech-centered investors to become wealthy. By 1994, the web first got to be accessible to the general public. Very quickly, companies saw the web as an important profit opportunity, America Online (AOL), Yahoo!, Amazon.com, EBay are a few examples of these opportunities. Technology stocks continued on taking off and made an exceptionally solid incentive for more technology startups to become traded at an open market. Tech startups kept on paying their workers in stock options, meant great benefits as long as stocks preserved their strong upward direction.

From 1996 to 2000, the NASDAQ stock index increased from 600 to 5000 points. "Dot-com" startups, went on running by individuals who were scarcely out of school, were opened up to the world and raised a huge amount of capital. A considerable number of these companies needed clear marketable strategies and significantly more had no profit at all.

By early 2000, a sense of reality began to return. Investors soon understood that the dot-com dream had degenerated into an exemplary speculative bubble. In the period of several months, the NASDAQ stock index slammed from 5000 to 2000, as shown in Fig. 7.4. Several stocks which once had a multi-billion dollar market capitalization, were off the map of the capital market as fast as they appeared. Panic selling resulted as the stock market's value broke down by trillions of dollars. The NASDAQ further decreased to 800 by 2002. At the same time, various bookkeeping scandals became known in which tech startups had falsely inflated their profits. In 2001, the U.S. economy encountered a post dot-com bubble recession, which forced the Federal Reserve to decrease interest rates again in order to stop the recession. A huge number of technology experts lost their occupations and, since they had invested into tech stocks, they also lost a noteworthy part of their life savings.

Fig. 7.4 NASDAQ composite index during the tech bubble (source: Thomson Reuters Eikon)

7.1.5 Subprime Mortgage Bubble

After the tech bubble and the September 11, 2001, the Federal Reserve stimulated a battling economy by decreasing interest rates to historically low levels. In under 2 years, from December 2000 until November 2002, the Fed decreased the rates from 6.5 to 1%. Contrasting this with the inflation rate, it is questionable that not taking on debt would infer losing cash. Accordingly, a housing bull market was created. Individuals with poor credit into this rush when mortgage lenders made non-conventional home loans: interest-only loans, installment options and home loans with augmented amortization periods.

Loan costs were generally low during the initial part of the decade. This low loan fee environment spurred increase in home loan financing and also in house prices. It urged investors to look for instruments that offer a yield enhancement. In this regards, subprime contracts offered higher yields than standard home loans and, thus, they have been in demand for securitization. The interest for progressively complex structured contract, for example, *collateralized obligation obligations* (*CDO*s) which embed leverage inside their structure, exposed investors to a more serious risk of default, however, with generally low financing costs and rising house prices. This risk was not viewed as excessive.

In the same time frame, financial markets have been abnormally liquid, which has cultivated higher leverage and more serious risk-taking. Spurred by enhanced risk management strategies and a movement by worldwide banks towards the supposed "originate-to-distribute" plan of action, where banks give loans, and then convey a great part of the credit risk arousing from these loans to end investors. Hence, financial innovation has prompted a sensational development in the business sector towards *credit risk transfer* instruments. In the course of the next 4 years, the worldwide outstanding amount of credit default swaps has increased more than ten times [7]. At that time, investors had a much more extensive scope of instruments available for them to price, repackage and circulate through the financial markets.

From the above discussions, it is obvious that if a small problem arose inside the subprime mortgage sector, it would rapidly spread to other sectors, too. In the long run, loan fees increased and numerous subprime borrowers defaulted when their loans' payments were updated into much more regularly scheduled installments. This left mortgage lenders with property that was worth less that the outstanding credit because of decreasing house prices. Defaults expanded; the issue snowballed, and some lenders went bankrupt. As subprime borrowers defaulted, investors in the subordinate tranche of the subprime CDOs took the first hit. This prompted lost confidence even among investors in the more secure tranches who had not endured any losses. But as they started to sell their investment, that panic started. The fire sale of assets led to a downward spiral of prices and a freeze in funding for these CDOs. In Fig. 7.5 we may see the real effect of this event on the *S&P* 500 index.

As we see from all of the popular bubble cases above, irrational expectations constantly generate damaging financial crashes. In fact, the enormous but enigmatically comprehended reality of bubbles requests a high excellence in academics. Nevertheless, academicians and national bankers attempt to discover a strategy to anticipate the financial bubbles and to build up a model and technique in order to forecast the bubbles.

In the upcoming section, we will identify how the *Illiquidity Premium* can be related to some financial bubble and how we can identify the early warning signs of the crisis with the help of *Illiquidity Premium*.

Fig. 7.5 *S&P 500* Index during subprime mortgage bubble (source: Thomson Reuters Eikon)

7.2 Illiquidity Premium vs. Financial Bubbles

Financial bubbles occur when the prices of assets transiently accelerate upward and raise over much their *fundamental value* (*intrinsic value*). The most common calculation method of the fundamental value of a stock or any security is the *discounted cash flow* (*DCF*) analysis that is defined as follows [3]:

$$\text{Fundamental Value of a stock} = DCF := \frac{CF_1}{(1+d)^1} + \frac{CF_2}{(1+d)^2} + \ldots + \frac{CF_n}{(1+d)^n}. \tag{7.1}$$

This formula benefits from *weighted average cost of capital* (*WACC*) as a discount variable to explain the time value of money. Here, d is a discount rate and CF_n $(n \in \mathbb{Z}_+)$ is a free cash flow along periods n, calculated in the following way [3]:

$$\text{Free cash flow} = \text{Net income} + \text{Amortization \& Depreciation} - \text{Changes in Working Capital} - \text{Capital Expenditures.}$$

In an efficient market, the price of an asset is equal to its fundamental value and, according to the fundamental value of a security, an investor can decide whether the security is *overvalued* or *undervalued*. If the intrinsic value of a security is higher than its current market price, it is classified as *overvalued*, if not, it is called as *undervalued*. If the price of an asset is less than its intrinsic value, a wise investor will purchase more to get profit. This behavior will boost the price of the security until no further profit can be obtained: in other words, until the price equals to the intrinsic value. If an asset is traded at a price which is higher than its fundamental value and if this case continue persistently, its price shows a bubble.

A bubble does not have a strong effect on the general economy if only a few investors are effected. Bubbles can lead to major troubles when they emerge in a financial instrument that is regularly held. Therefore, devastating bubbles are, for their major parts, the one in securities' exchanges [5]. Since stock market bubbles are only realized during a continuous bull market circumstance, the confidence of the traders is very high [6]. They believe that the demand for the stocks will never end and stocks will always become profitable. This belief about the stock market causes irrational expectations, and it escalates the stock prices upward and inflates the size of the stock market bubble as well. This trend ends when some investors recognize that the prices have risen unrealistically, thus they begin selling their stocks before the prices go down. Then, other traders follow this attitude; hence, panic selling starts. Not always but mostly, this process is completed by a sharp decline and when this acute drop occurs, it is said that the *bubble bursted*.

Most of the firms that are growing more and more rapidly through a stock market bubble go bankrupt when the bubble bursts. This gives rise to an increase of the unemployment rates [6]. Business and consumer consumptions diminish and this may cause to commence an economic recession [6]. Because of these negative effects of bursting of a bubble, it is important to develop *early-warning* methods to detect it, timely [19]. However, it is still exceptionally hard to describe, to calculate and to keep them from further inflating ahead of time. Therefore, one of the various scientific contributions of this thesis is to introduce an early-warning method for bubbles, for the first time by using derivative markets as an application field of Conic Finance theory which is a very new field in the literature. This theory is deeply connected with liquidity effects and risk behavior of financial markets.

Here, as a new approach, we use implied liquidity from Conic Finance theory. In financial markets, liquidity is an important feature. It mirrors the asset's capability to be purchased or sold without a huge change in the price and with a least loss of value. Liquidity is closely associated with bid-ask spread: highly liquid products have a narrow spread; illiquid assets have a wide spread. During financial shocks, the liquidity is said to evaporate and, hence, drive further the bid and ask prices of financial securities from their fundamental values. Therefore, implied liquidity, or as we called it in previous chapters, *Illiquidity Premium* has been thought as a kind of indicator to detect bubbles. Also, Conic Finance theory is founded on the basis of two concepts: acceptability and distorted expectations. The bubbles can be interpreted as distorted expectations of investors; therefore, we benefited from this theory to identify stock market bubbles.

In Chaps. 4 and 6, we have already calculated the daily illiquidity premium for the derivative market, and we saw that the premium increases during the times of disturbance or recession and it decreases during the times of recovery and expansion. In this regard, as we have mentioned in preceding paragraphs, illiquidity premium can be regarded as an indicator for the market contraction or recession and, specifically, for financial bubbles. To do this we are going to utilize the *sliding window* technique.

It is common practice to use a sliding-window to dynamically model the changing properties, in our case: the illiquidity premium, of a single or multiple time series. The simplest method makes use of a window of a fixed length which slides through the data at fixed intervals, usually one datum at each step. This involves updating some or all of the model parameters at each time step, using a window in which a fixed number of past data points are used to estimate the parameters. We now describe some major criteria which need to be considered when selecting the appropriate window length for a fixed-length sliding-window. The window should be large enough so as to accurately capture any variation of the market signals within it. However, a large window also may not be able to properly reflect rapid changes in the dynamics of the market and may result in big computational times. Therefore, the window should be small enough so as to accurately compare disparity of the market signals at corresponding points [26], without leading to "noisy" results. However, a very small window may not contain enough data points for an accurate computation of the dependency measure, which is illiquidity premium. Hence, any sliding window model carries this complexity-benefit trade-off with respect to the choice of window length.

Sliding window analysis of time series is usually applied to dynamically update the parameters of a model. It is a common form to compute the parameter estimates through a window of sample data which is sliding in a fixed length. The estimates of model parameters over the sliding windows should be similar if the data are stationary. On the other hand, if the parameters change frequently or at some observations, the rolling window analysis is able to detect the unstable changes over the estimations [29].

In this thesis, we concentrate on dynamic variation of the stock markets' movements advancing with time t, so we examine the illiquidity premium estimated over a sliding window. In order to do this, we create a time-evolving sequence of matrices by rolling the time window of τ prices through the full data set. Saying in another way, τ will represent the sliding window length.

As having been indicated by us in the above paragraphs, the choice of τ is a compromise between an excessively noisy and an over excessively smoothed curve [27]. Also, it must be taken in consideration the type of data that we are dealing with. In this work, it would be interesting to study sizes τ of the rolling window to be $\tau = 5$ or $\tau = 22$ trading days, i.e., weekly and monthly data. Generally, the shorter the interval, the earlier the *early-warning* for the crises and the better the monitoring of the price process [18].

Equations (6.26), (6.27) or (6.28) are applied to calculate the illiquidity premium over a subset of option bid and ask price series within the rolling window $[t-\tau+1, t]$. For instance, the illiquidity premium in the first sliding windows are computed by the price series within $[1, \tau]$ and $[2, \tau+1]$ for the following sliding window. By only shifting the time window by one data point, there is a significant overlap in the data contained in consecutive windows. This approach enables us to track the evolution of the derivative market's illiquidity premium and to identify time steps at which there were significant changes in it.

As we have noted, a sliding window approach will be used to analyse and calculate the values for the illiquidity premium with respect to the data set. This will help us to confine the search for "*early warning signs*" to a few windows before and after the events of interest. It is in the nature of this approach that we can apply this technique to different intervals of fixed size. Each one of these intervals can be characterized by different results. Another purpose of our analysis on different scales is to test the dependence of the results on the granularity of the data, since we expected different behaviors at different scales for financial time series. The other studied criterion is the window size. We wonder: Do the results, in general, remain the same independently of the size of the window?

Let us remember that in Eqs. (6.26)–(6.28), the parameter τ stood for the number of days for which we want to calculate the illiquidity premium. As we have mentioned, for the weekly premiums we have to take $\tau = 5$, as the number of business days in a week is five. Under these conditions, minimizing TSE gives the market level's weekly γ. Moreover, we generate sliding window values of γ by choosing a windows of 5 days of length and sliding it along the historical data, beginning at the first date (January 2, 2008) and continuing until the window reaches the last date of the data (December 12, 2010). The window moves forward in steps

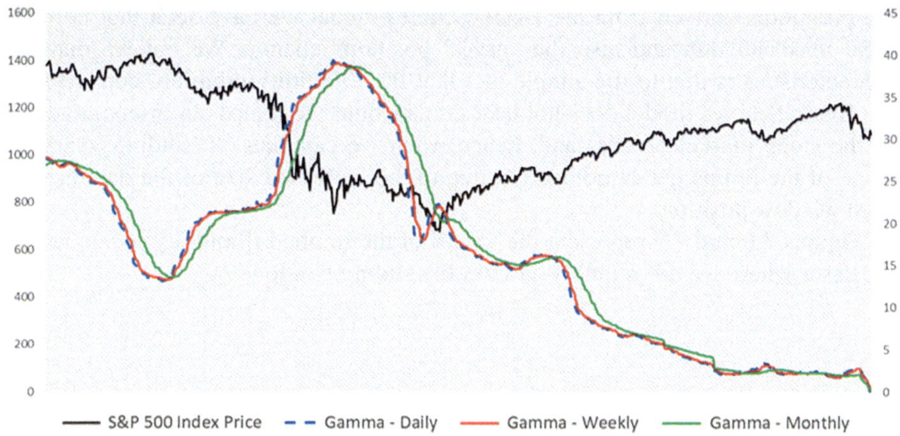

Fig. 7.6 Illiquidity premium derived from the Kou model

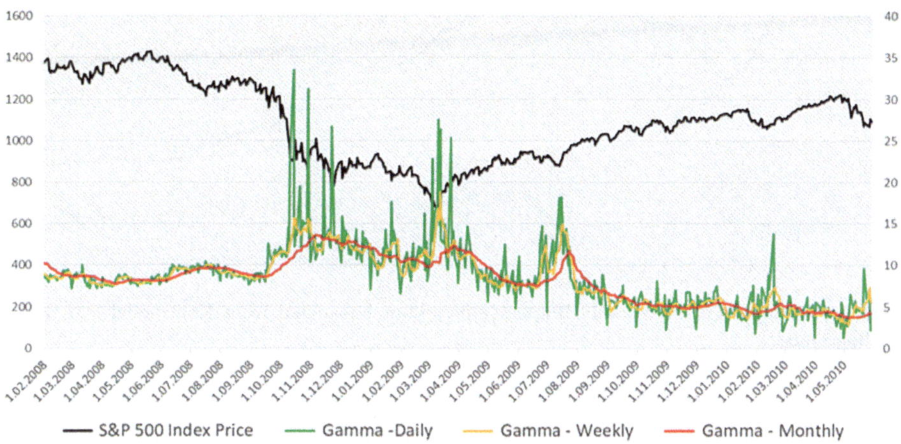

Fig. 7.7 Illiquidity premium derived from the Black-Scholes model

of 1 day. At each position of the window, the value of γ is calculated, by using the total sum-of-squares method, and assigned it against the window position.

We did the same calculation steps for the monthly values, too, but in this case we took 22 working days. The results of these calculations can be found in Fig. 7.6 for the Kou Model and in Fig. 7.7 for the Black-Scholes Model: for a comparison we also provide the results of the daily illiquidity premium curve.

By analyzing the above figures we may see that the γ never takes zero values, as this indicates that in the two-price world the bid and ask prices never equal to each other. One main difference between the illiquidity premium derived from Kou and Black-Scholes models are the trend that the different curves follow. In premiums derived from the Kou model, we may see that the curves do not have a significant change in local max and min values; they only shift to the right. Meanwhile, in the premiums derived from the Black-Scholes model we have seen that both the local max and min and also the curves' positions change. We believe that this characteristics is due to the simple fact that the illiquidity premium derived from the Black-Scholes model does not take into account the jumps, an inseparable part of the stock market process and, hence when we calculate the sliding windows, some of the jumps get smoothed out due to the increased size of the data used in each window position.

Figures 7.8 and 7.9 represent the values of the implied illiquidity parameter for the cases where we use monthly and weekly sliding windows.

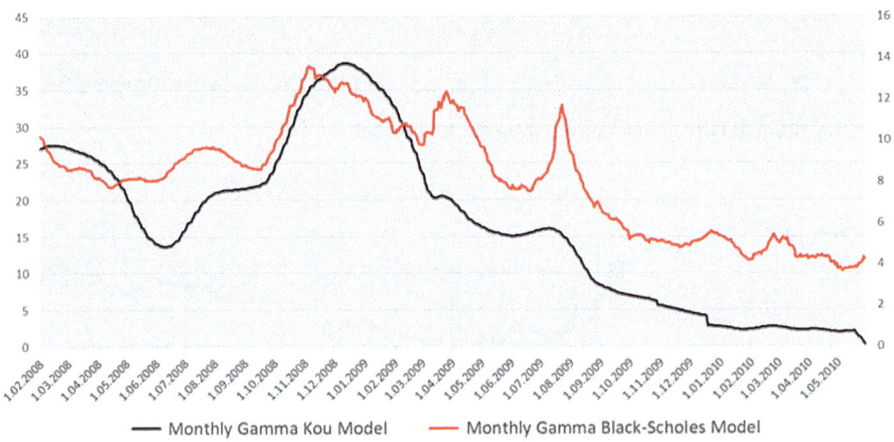

Fig. 7.8 Comparison of monthly illiquidity premiums calculated from Kou model and Black-Scholes model

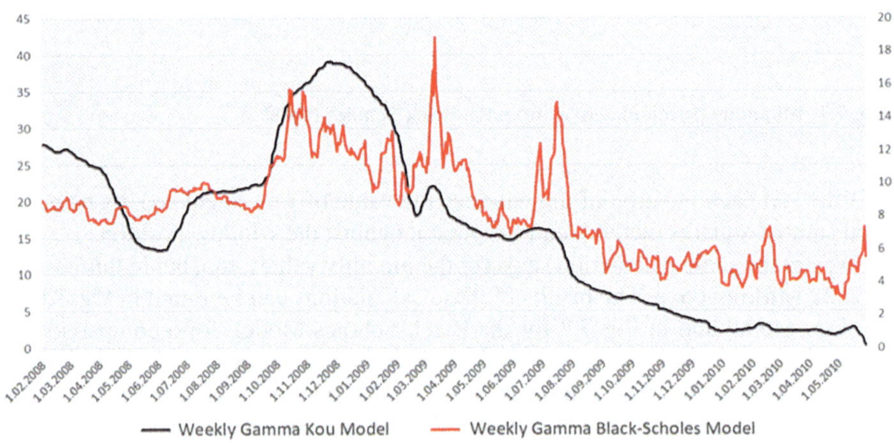

Fig. 7.9 Comparison of weekly illiquidity premiums calculated from Kou Model and Black-Scholes model

As a result, we observed a sharply increasing process in the illiquidity premium obtained from the derivatives market, when the bubble-burst time approaches. Moreover, the increase in the illiquidity premium over periods, like financial trouble of Chrysler and General Motors or European debt crisis, etc., is lower than the increase which we have calculated for the financial crisis and the bubble-burst time, which means that their effects on the U.S. economy were not as critical as the Subprime Crisis.

As final remarks of this chapter, we may deduct that an increase in the illiquidity premium goes hand in hand with problems in the financial market and, in particular, with the financial crises. In our opinion, this is related with the investors' unwillingness to execute the order during uncertain times, which can be the times when there is an insufficient information regarding the current economical conditions or during the deteriorating state of the systematic institution, as in the cases of Chrysler/General Motors or European debt crisis.

7.3 Comparison with Other Bubble-Detection Techniques

Bubbles are still a controversial subject in classical economics; they imply that there is sizable and persistent deviation between the fundamental value of an asset and its market value. But the dominant paradigm in economics, the Efficient Market Hypothesis (EMH) [22], states that all information about the fundamentals of an asset are reflected in the market price through the action of the rational market participants. As such, the fundamental and the market value are the same: if there was a difference between the two, there would be an arbitrage opportunity and that difference would quickly be traded away. An important challenge to this paradigm is the empirical evidence about bubbles, such as the historical examples cited at the beginning of Sect. 7.1. Furthermore, while economists who base their assumptions on EMH argue that even large bubbles, such as the Tulip Mania, can be explained fundamentally, a growing number of models have been proposed to explain these phenomena. The theories on bubbles can be classified into three broad categories: rational bubbles, heterogenous belief bubbles and behavioral bubbles. Before giving an overview of the main bubble-related literature, we will explain the crucial concept of (fundamental) value in economics. This overview of the existing literature does not aim to be exhaustive, but rather to give a basic understanding of some of the main theories.

7.3.1 Value in Economics

The notion of value in economics is based on two fundamental concepts. The first one can be enunciated as follows: \$100 today is worth more than \$100 tomorrow. Generalizing this elementary concept, the present value of \$$CF$ in t years with an annual return of d is:

$$\text{Present Value} = \frac{CF}{(1+d)^t}, \tag{7.2}$$

where the exponent t comes from compounding. As we have indicated in Eq. (7.1) the value of an asset distributing dividends CF_t every year is:

$$\text{Fundamental Value (Price)} = \sum_{t=0}^{\infty} \frac{CF_t}{(1+d)^t}. \tag{7.3}$$

The second fundamental concept relates to the magnitude of d, also called *risk premium* or *weighted average cost of capital*. Economics postulate a positive relationship between risk and return. In other words, a higher risk has to be remunerated by a higher return. The value d has a lower bound called the risk-free rate (r_f) which is the remuneration that one gets for a riskless investment. It embodies time value and for the U.S. market is usually proxied by the return on U.S. T-bills. Determining the value of the additional component of d associated with the riskiness of an asset is the major problem of economics. Theories such as the Capital Asset Pricing Model [33] or the 3-factor model [11] propose specific relationships between risk and return. Although this economic definition of value is very narrow, since it ignores other dimensions of value such as the social or tactical dimensions, it offers a well-defined framework to postulate and test hypotheses.

7.3.2 *Rational Bubbles*

Rational bubbles are probably the dominant approach taken by economists to explain the emergence of bubbles. These models examine the conditions under which bubbles can form given that all the agents are rational.

The foundation of the rational bubble models is the one proposed by Blanchard and Watson [4]. In this model, the value of an asset is decomposing into its fundamental value and a bubble component:

$$\text{Price}_t = \text{Price}_t^{\text{fund}} + B_t = E_t\left[\sum_{\tau=t+1}^{\infty} \frac{CF_t}{(1+d)^{\tau-1}}\right] + \lim_{T\to\infty} E_t\left[\frac{B_T}{(1+d)^{T-t}}\right], \tag{7.4}$$

where the fundamental component is the present value of the discounted future cash-flows (cf. Eq. (7.3)) and the bubble component is the present value at $T \to \infty$. The first consequence of this model is that, for the bubble to exist, it has to grow with rate d. Indeed, if the bubble would grow with growth rate $d_B < d$, the present value of the bubble component would be 0 and, as such, the price of the asset would be equal to its fundamental value. In this case, we will have to replace B_T with $B_0(1+d_B)^{T-t}$ in Eq. (7.4). If on the contrary, $d_B > d$, the bubble component would be infinite and, as a consequence, so would be the price. The second implication of the model is that the asset has to be lived infinitely: suppose that the asset has a maturity date T. At that time, the asset would be liquidated for its fundamental value and $B_T = 0$. But if $B_T = 0$, nobody would be willing to buy the asset for more than its fundamental value at $T-1$, knowing that one time step later, the bubble component will be 0. This backward induction argument implies that a bubble can only exist for infinitely lived assets.

One of the many problems of the Blanchard and Watson's model is that, as the bubble component of the price grows exponentially, the price-to-cash-flow ratio becomes infinite, i.e., $\lim_{T\to\infty}(\text{Price}_T/CF_T) = \infty$, which is unrealistic. To remedy this problem, Froot and Obstfeld [13] proposed to make the bubble component depend on the cash flows rather than on time. This choice was motivated by the observation that investors are bad at predicting future cash flows. By doing so, the authors show that under the hypothesis of no bubbles, the price-to-cash-flow ratio would be constant, $\text{Price}_T/CF_T = k$, with k being a constant. Applying this criteria on the *S&P 500* over the 1900–1988 period, they rejected the hypothesis of no bubbles.

By relaxing the assumption about common knowledge, i.e., the fact that everybody knows that everybody knows, and limiting short-selling, Allen et al. [2] show that bubbles can form on finitely lived assets. The intuition behind this result is that the absence of common knowledge eliminates the backward induction argument, since a rational agent can hope to resell the asset to another one who might not know. Furthermore, constraining short-selling limits the ability of the agents to learn other agents' private information from market prices.

Another class of models departs from the assumption that all agents are rational and achieves bubbles by introducing a second class of traders that are behavioral feedback traders. It is the interaction between the rational arbitrageurs and the behavioral traders that creates the bubbles. Delong et al. [10] show that, in their setup, rational arbitrageurs buy the asset after a good news in order to bait the behavioral feedback traders into pushing the price further up, allowing the rational agents to sell their stock shares profitably at the expense of the behavioral ones.

In Abreu and Brunnermeier [1], all the rational agents know that there is a bubble, but it is the difference of opinions about the timing of the start of the bubble or, in other words, their estimate of the asset's fundamental value that leads to a synchronization problem, each agent not being able to burst the bubble on its own. As such, rational arbitrageurs ride the bubble until they can synchronize, due to a piece of exogenous information for example. These models offer a powerful argument against the EHM which claims that even if there are irrational agents in the market, rational arbitrageurs will prevent any possibility of mispricing.

Lin and Sornette [20] on the other hand argue that it is the difference of opinions about the timing of the end of the bubble leading to the persistence of bubbles. They support their claim by applying their model on real data and derive an operational procedure that allows them to diagnose bubbles and forecast their termination.

7.3.3 Heterogeneous Beliefs Bubbles

Heterogeneous belief bubbles take place in a setup where agents do not agree on the fundamental value of the asset. This can be due to psychological biases, or due to the difficulty in making predictions about an uncertain future. Miller [24] shows in a very simple framework that given a limitation on short selling, the divergence in opinion between the agents about the asset's return leads to an equilibrium where the price is higher than the average estimate. Put simply, the optimists push the price higher than the average estimate because the pessimists stay out of the market, not being able to fully reflect their opinion by shorting the asset. Moreover, the author shows that the price of the asset increases with the diversity of the opinions about its future returns.

In a dynamic framework, Harrison and Kreps [14] demonstrates that not only can a bubble arise when agents have different opinions about the fundamentals of an asset, but the price of the asset can even surpass the valuation of the most optimistic agent. This happens because the optimistic agent chooses to pay a premium to buy the asset in the hope of reselling it later when he will be pessimistic (and other agents will be optimistic). We should note that short-selling is also restricted in this model.

Scheinkman and Xiong [32] build their model on Harrison and Kreps [14] model and extend it into continuous time. They interestingly conclude that bubbles are characterized by higher trading volumes, a fact that can be observed empirically.

7.3.4 Behavioral Bubbles

Some financial economists agree that psychological biases must play a fundamental role in the formation of bubbles, departing completely from the claim that agents are rational, and frontally attacking the EHM. Shiller [34] cites several behavioral mechanisms to be at the origin of bubbles. Among the most relevant ones are positive feedback loops between price and investor's enthusiasm, as well as herding, i.e., the fact that people tend to imitate each other. In the light of the historical bubbles and crashes, some of which were described in previous sections, these mechanisms seem very convincing in generating bubbles. It is also worth mentioning the work of Hens and Schenk-Hoppé [15] who show in an evolutionary framework, where strategies implemented by heterogenous agents with different opinions and behaviors compete for market capital, that seemingly irrational strategies can outperform seemingly rational ones.

Models coming from physics, in particular the Ising model, have been very successful at describing how imitation between agents can lead to bifurcation in their aggregate opinions [35]. The Ising model consists of agents influencing each other. In a nutshell, if an agent is surrounded by agents willing to sell, it is likely that he or she will start selling as well. Two opposite forces are at play: the ordering force of social imitation and the disordering force of idiosyncratic noise. In an Ising-like framework, Orléan [28] shows that Bayesian opinion leads to two qualitatively different dynamics. When the agents' estimates of the group's opinion, i.e., the estimation of the fraction of agents that would buy, are heterogenous, the stationary distribution of opinions is peaked around 50%: half of the agents would buy and the other half would sell. However, when the same estimate is homogenous, the interaction among agents leads to a stationary distribution with two peaks: most agents would buy and a few would sell, or most agents would sell and a few would buy. These situations can be interpreted as bubble and crash states, respectively.

Lux [21] sets up a framework with two kinds of agents: speculative traders and fundamentalists. Speculative traders form their decisions based on the opinion about other traders of their kind, as well as on price dynamics (momentum). Fundamentalists buy or sell based on the difference between the asset's market and fundamental values. The interaction between the Ising-like speculative traders and fundamentalists gives rise to a rich phenomenology of price dynamics, with prices moving around an equilibrium as well as boom and bust cycles, depending on different parameters. This shows that simple mechanisms are enough to generate a variety of dynamical regimes.

Contrary to the previous works, where every agent was interacting with every other, Cont and Bouchaud [9] impose a random graph topology on how the agents interact. A random graph is a network where an agent has a probability c/N to be connected to another agent, N being the total number of agents and c the average number of connections of per agent. In their setup, the authors propose that every agent of a component, i.e., the set of agents connected through a path, would take the same action of buying, selling or staying out of the market. The components could be thought of as organizations such as hedge funds, individual traders, etc. The main result of the paper is that for $0 < c < 1$, the fat-tails of the distribution of returns can be recovered. The case $c = 1$ corresponds to a critical value where a giant component emerges, encompassing a finite fraction of the system. This can be interpreted as a bubble or a crash, since the giant component contains agents with the same action.

Although nearly all of the models described so far offer an explanation as to what mechanisms could be at the origin of the formation of bubbles, they suffer from a major limitation: either they cannot be calibrated to real data and as such are not testable, or they lack any predictive power. Hence, among the different types of the bubble models that we have described above, the model that can be related to the model developed in this thesis is Log-Periodic Power Law (LPPL) model. We are going to give a comparison between the LPPL model and Conic Finance model (or Illiquidity Premium model) in the subsequent section.

7.4 Log-Periodic Power Law Model vs. Illiquidity Premium

The major model that distinguishes itself by proposing a functional form for the price dynamics up to the crash is the LPPL model and was first suggested by Sornette et al. [37]; it was later formalized by Johansen et al. [17]. LPPL model is constructed on the behavioral and well-documented phenomenon of positive feedback between demand and price, i.e., the fact that an increase in price leads to an increased demand in speculative periods, which itself leads to an increase in price. Translating the positive feedback mechanism into mathematical terms, one can show that the price dynamics obeys a super-exponential law with a finite-time singularity, beyond which the solution does not exist. This finite-time singularity can be interpreted as the time of the crash. In addition, imposing a hierarchical topology on the networks of traders leads to the super-exponential price dynamics, being characterized by log-periodic oscillations of decreasing amplitude. Other explanations for the log-periodic oscillations include the rivalry within nonlinear trend followers and nonlinear value investors [16].

The LPPL model was expanded to interpret the dynamics of financial markets during bubbles and crashes. Similar to the agent-based model it is assumed that there are rational traders and noise traders who display herding behavior that can destabilize the asset price [12]. Employing the LPPL approach, Sornette et al. [37] have analyzed the stock market bubbles and crashes at the macroeconomic and microeconomic levels. From a macroeconomic perspective, the model presumes that we are dealing with rational markets which have incomplete information. In this kind of market, the trade price will reflects both the fundamental value and also the future expectations associated with the profitability and risk. From a microeconomic perspective, the Sornette-Johansen model considers that investors, both rational investors and noisy traders, are connected narrowly through the networks that govern their anticipations regarding market earnings. Moreover, trade choices rely upon the choices of other members of the network, but may also consist of external influences.

One of main disadvantage of LPPL model is that it can not differentiate between several sources of price growth as described below:

- Possibility that *the prices are high and keep increasing as investors expect high future earnings growth.* This growth could be related to a shift in the economic structure and the way economic gains are divided between companies and employees, or between retaining earnings and dividend payments.
- Possibility that *there could be a bubble in the sense that prices are high today, simple because investors anticipate future prices to be even higher, without taking into account the fundamentals.*
- Possibility that *bubbles and crashes can be formed by fluctuations in investors' risk premia.* Sudden rises in prices can be result of the investors suddenly becoming more risk tolerant and requiring a lower expected return for the asset. Contrary, as risk premia suddenly change its course, markets can experience crashes. Prices in accordance with this theory are always rational, and reflect

investors' stand towards risk. One might see a sharp increase in asset prices, but one should not clarify this as inconsistent with the fundamentals [30].

Another disadvantage of the LPPL model is that for pre-crash bubbles on stock markets the LPPL model [17] require the expected prices to be non-decreasing during the whole bubble period (as recognized by Sornette and Zhou [36]). In the same study, it was described that the LPPL model fitted to the price process decreases at some point during the bubble.

Moreover, estimating LPPL models in general never was an easy task, as there are numerous local minimum of the price function where the minimization method can get trapped.

However, in a series of papers, Madan, Cherny and their co-authors developed a new method which succeeded to overcome the shortcomings of the aforementioned traditional approach.

Conic Finance is a new way to price financial assets; it combines No-Arbitrage theory and Expected Utility theory. The level of acceptability is very important in Conic Finance, because it helps market-makers to set good bid and ask prices. In general, this theory is very helpful for investors and portfolio optimizers to distinguish between different investment opportunities. It disregards the law of one-price by incorporating the bid and ask prices. Moreover, by being established on the basis of two principles, namely, Acceptability and Distorted Expectations, Conic Finance theory is deeply connected with liquidity effects and risk behavior within financial markets.

We can empirically characterize bubbles as periods in which illiquidity premium experiences explosive dynamics, and produce statistical tools to discover and time-stamp the existence of such a behavior.

When applied to historical bid and ask prices of the European options written on the *S&P 500* index, the method classifies five cases where bid and ask prices deviated from fundamentals and in which there were elevated levels of uncertainty in the market between the start of the 2008 and end of 2010.

Subsequently, we summarized the advantages of an Illiquidity Premium model:

- An Illiquidity Premium model does not require to differentiate between the different sources, different types of price growth.
- An Illiquidity Premium model determines a real-time tool for identification of the bubbles. To decide whether an asset is in the state of a bubble or not, only current and past information is necessary and there is no look-ahead bias.
- An Illiquidity Premium model has no constrains on prices to be increasing throughout the bubble.
- Instead of providing an answer to the question of if there is a bubbles in asset prices, Illiquidity Premium model specifies a live-dating algorithm in order to identify when bubble end, which can be beneficial to the stock market investors.
- An Illiquidity Premium model quantifies the bubble state of an asset through its price dynamics, without resorting to the determination of a fundamental value.

- The increase in the illiquidity premium in this model leads to a critical point that characterizes the beginning of the market crash.

Yet these models have never been compared to each other in a statistical meaning. The major reason is that the data used in LPPL model are the stock prices, or *S&P 500* index prices, but the data that is used in the Illiquidity Premium model are the bid and ask prices of the European options.

7.5 Investment Management and Illiquidity Premium

This chapter has defined a new mathematical approach for detecting the termination of bubbles in real-time. The technique does not have the disadvantages of traditional methods, and demonstrates that bubbles and consequent price increases of the market are a frequent characteristics of stock market.

For investors and investment managers, being able to determine periods when asset prices are in a bubble state in real-time and when this bubble bursts has crucial implications for portfolio optimization and hedging. Moreover, correctly determining the periods of enormous exuberance and panic is valuable for measuring how different strategies function in these episodes with respect to more calm periods of market behavior.

We choose as default strategy for non-bubble times that the investors are rational in the current model, i.e., investing a fixed percentage in the risky asset. An investor will consider that an economy is experiencing economic uncertainty or, particularly, a financial bubble as soon as the illiquidity premium of the bid and ask prices derived from the model in consecutive time windows will yield increasing results. If investors detect this signal, i.e., increased illiquidity premium, they decrease their investment in the risky asset to 0% of their wealth to limit the loss arising from the bubble bursts. As soon as the signal is not picked up anymore, i.e., the illiquidity premium is gradually decreasing or is stable, investors will have to reevaluate their strategy and revert to their previous behavior of investing a fixed percentage of their wealth in the risky asset.

References

1. D. Abreu, M.K. Brunnermeier, Bubbles and crashes. Econometrica **71**(1), 173–204 (2003)
2. F. Allen, S. Morris, A. Postlewaite, Finite bubbles with short sale constraints and asymmetric information. J. Econ. Theory **61**(2), 206–229 (1993)
3. S. Alvarez, What is the intrinsic value of a stock? (2013), http://tinyurl.com/nc5znh7
4. O.J. Blanchard, M.W. Watson, Bubbles, rational expectations and financial markets (1982)
5. J. Calverley, *When Bubbles Burst: Surviving the Financial Fallout* (Nicholas Brealey Publishing, Boston, MA, 2011)
6. S.-L. Carty, Definition of stock market bubble (2013), http://tinyurl.com/zf3r2qo

7. A. Cates, Global economic watch, UBS Investment Research, Global Economics Research (2008)
8. J. Colombo, The stock market crash of 1929 (2012), http://www.thebubblebubble.com/1929-crash/
9. R. Cont, J.-P. Bouchaud, Herd behavior and aggregate fluctuations in financial markets. Macroecon. Dyn. **4**(02), 170–196 (2000)
10. J.B. De Long, A. Shleifer, L.H. Summers, R.J. Waldmann, Noise trader risk in financial markets. J. Polit. Econ. **98**, 703–738 (1990)
11. E.F. Fama, K.R. French, The cross-section of expected stock returns. J. Financ. **47**(2), 427–465 (1992)
12. V. Filimonov, D. Sornette, A stable and robust calibration scheme of the log-periodic power law model. Phys. A Stat. Mech. Appl. **392**(17), 3698–3707 (2013)
13. K.A. Froot, M. Obstfeld, Intrinsic bubbles: the case of stock prices, Technical report, National Bureau of Economic Research (1989)
14. J.M. Harrison, D.M. Kreps, Speculative investor behavior in a stock market with heterogeneous expectations. Q. J. Econ. **92**, 323–336 (1978)
15. T. Hens, K.R. Schenk-Hoppé, Survival of the fittest on wall street, Discussion Paper, 2004
16. K. Ide, D. Sornette, Oscillatory finite-time singularities in finance, population and rupture. Phys. A Stat. Mech. Appl. **307**(1), 63–106 (2002)
17. A. Johansen, O. Ledoit, D. Sornette, Crashes as critical points. Int. J. Theor. Appl. Fin. **3**(02), 219–255 (2000)
18. E. Kürüm, Early warning on stock market bubbles via ellipsoidal clustering and inverse problems, Ph.D. thesis, Program of Financial Mathematics, Institute of Applied Mathematics, Middle East Technical University, Ankara, 2014
19. E. Kürüm, G.-W. Weber, C. İyigün, Financial bubbles, in *Modeling, Dynamics, Optimization and Bioeconomics I* (Springer, Cham, 2014), pp. 453–468
20. L. Lin, D. Sornette, Diagnostics of rational expectation financial bubbles with stochastic mean-reverting termination times. Eur. J. Financ. **19**(5), 344–365 (2013)
21. T. Lux, Herd behaviour, bubbles and crashes. Econ. J. **105**, 881–896 (1995)
22. B.G. Malkiel, E.F. Fama, Efficient capital markets: a review of theory and empirical work. J. Financ. **25**(2), 383–417 (1970)
23. J. Marwood, 10 biggest stock market bubbles and crashes of all time (2015), http://jbmarwood.com/bubbles-crashes/
24. E.M. Miller, Risk, uncertainty, and divergence of opinion. J. Financ. **32**(4), 1151–1168 (1977)
25. H.P.H. Nusteling, *Welvaart en werkgelegenheid in Amsterdam, 1540–1860: een relaas over demografie, economie en sociale politiek van een wereldstad* (Bataafsche Leeuw, Amsterdam, 1985)
26. M. Okutomi, T. Kanade, A locally adaptive window for signal matching. Int. J. Comput. Vis. **7**(2), 143–162 (1992)
27. J.-P. Onnela, A. Chakraborti, K. Kaski, J. Kertesz, A. Kanto, Dynamics of market correlations: taxonomy and portfolio analysis. Phys. Rev. E **68**(5), 056110 (2003)
28. A. Orléan, Mimetic contagion and speculative bubbles. Theor. Decis. **27**(1–2), 63–92 (1989)
29. D. Pequeno et al., Exchange rate changes and net positions of speculators in the EuroFX futures market-does market size matter? Department of Economics, Aalto University School of Economics, 2011
30. P.C. Phillips, J. Yu, Dating the timeline of financial bubbles during the subprime crisis. Quant. Econ. **2**(3), 455–491 (2011)
31. G. Richardson, A. Komai, M. Gou, D. Park, Stock market crash of 1929 (2013), http://www.federalreservehistory.org/Events/DetailView/74
32. J.A. Scheinkman, W. Xiong, Overconfidence and speculative bubbles. J. Polit. Econ. **111**(6), 1183–1220 (2003)
33. W.F. Sharpe, A simplified model for portfolio analysis. Manag. Sci. **9**(2), 277–293 (1963)
34. J. Shiller Robert, Bubbles, human judgment, and expert opinion. Financ. Anal. J. **58**(3), 18–26 (2002)

35. D. Sornette, Physics and financial economics (1776–2014): puzzles, Ising and agent-based models. Rep. Prog. Phys. **77**(6), 062001 (2014)
36. D. Sornette, W.-X. Zhou, Predictability of large future changes in major financial indices. Int. J. Forecast. **22**(1), 153–168 (2006)
37. D. Sornette, A. Johansen, J.-P. Bouchaud, Stock market crashes, precursors and replicas. J. Phys. I **6**(1), 167–175 (1996)
38. M.C. Taylor, *Confidence Games: Money and Markets in a World Without Redemption* (University of Chicago Press, Chicago, 2008)
39. E.A. Thompson, The tulipmania: fact or artifact? Public Choice **130**(1–2), 99–114 (2007)

Chapter 8
Conclusion and Future Outlook

In our thesis, we present a new measure to account for the differences between the bid and ask prices of the financial instruments and an early-warning signaling for financial bubbles by benefiting from Conic Finance theory, modern optimization techniques and well-chosen numerical methods. For the time being, we consider the U.S. derivative markets since bubbles, mostly, occurred in these markets. Yet, we intend to extend our work for further countries' markets and also for other types of assets, such as gold and oil markets.

Conic Finance is a new way to price financial assets: it combines the No-Arbitrage theory and Expected Utility theory. The level of acceptability is very important in Conic Finance, because it helps market-makers to set good bid and ask prices. In general this theory is very helpful for investors and portfolio optimizers to distinguish between different investment opportunities. It disregards the law of one-price by incorporating the bid and ask prices. Moreover, by being established on the basis of two principles, namely, Acceptability and Distorted Expectations, Conic Finance theory is deeply connected with liquidity effects and risk behavior within financial markets.

In markets of the financial sector, liquidity is a core criterion. It mirrors the asset's capability to be bought or sold without a significant change in the price and with a lowest loss in value. Liquidity is closely associated with bid-ask spreads: highly liquid products have narrow spreads; illiquid assets have a wide spread. During financial shocks, liquidity is sometimes said to evaporate. Therefore, it is increasing the bid and ask prices of financial securities. Financial bubbles can be interpreted as distorted expectations of investors; hence, we can benefit from Conic Finance theory in order to recognize and identify the stock market bubbles as early as possible. In this thesis, in line with the numerical calculation of the illiquidity premium, we presented a new approach to identify stock market bubbles by benefiting from Conic Finance theory, and illiquidity premium was understood to be a kind of indicator to detect bubbles.

A. Karimov, *Identifying Stock Market Bubbles*, Contributions to Management Science, https://doi.org/10.1007/978-3-319-65009-8_8

In this thesis we deal with the 2008 Subprime Crisis in equity markets, by analyzing the derivative markets and the movement of the acceptability index of European options written on the *S&P* 500 index during that financial crisis, from 2008 and December 2010.

In order to calculate the illiquidity premium we firstly employ the Black-Scholes Model and extend it so that it will include our implied liquidity parameter. By this way we derive the bid and ask prices of the European put and call options. The second model that we use is the Kou model. This model have been chosen as it better represents the movements in the stock markets. In this case, by using the Inversion formula for densities, Residue Theorem from Complex Analysis, Taylor Series Expansion and from Theory of Inverse Problems we also derive the bid and ask prices of the European call and put prices that incorporate the illiquidity premium.

Moreover, in order to monitor the market movements closely, weekly and monthly sliding windows have been used. As a result, we observed a sharply increasing process in illiquidity premium obtained from the derivatives market, when the bubble-burst time approaches. Moreover, the increase in the illiquidity premium for other periods, like financial trouble of Chrysler and General Motors or European debt crisis, etc., is lower than the increase which we have calculated for the financial crisis and its bubble-burst time.

Conic Finance theory is a very new field in the finance literature and its applications have begun very recently. One of the various scientific contributions of this thesis is the introduction of an early-warning method for bubbles, for the first time by using derivative markets as an application field of Conic Finance theory which is a very new field in the literature. In this regards, we can empirically characterize bubbles as periods in which the illiquidity premium experiences explosive dynamics, and produce statistical tools to discover and time-stamp the existence of such a behavior.

We are going to extend our work to derivatives contracts which are written on other countries' stock markets and also to other types of assets such as gold and petrol markets. This thesis analyzes the derivatives market's illiquidity based on daily data; it would also be very challenging and interesting to take a look at how illiquidity premiums will behave with respect to higher frequency data, for instance, at a trade-by-trade frequency. What is more, we shall address financial dynamics beyond the use of geometric Brownian motion or Kou Model and investigate the application of more advanced processes in the calculation of implied liquidity or illiquidity premium. Moreover, we are going to apply a more complex approach for the window-size determination by using adaptive windows whose sizes change with time depending on some properties of the underlying assets or the entire economy. This will offer a possible compromise between the two conflicting criteria, namely, accurately capturing the variation of the signals and accurately comparing disparity of the signals at corresponding points, for selecting the windows' lengths.

Eventually, we aim at practical tools and a Graphical User Interface for the analyst, manager and decision maker, related with our computational and probabilistic methods on the detection, early warning of financial bubbles and further problems of liquidity. In addition, we will be working on a more detailed investment portfolio management strategy that will use illiquidity premium as a signal to change the combination of different investments.

Appendix A
Some Distributions Under Wang Transform

Assume that the function Ψ^{γ} is the Wang transform, then it will have the following formula:

$$\Psi^{\gamma}(u) = \Phi(\Phi^{-1}(u) + \gamma), \quad u \in [0, 1], \quad \gamma \geq 0, \tag{A.1}$$

where Φ is the standard normal distribution function.

Wang transform of the normal and lognormal distributions are also normal and lognormal distributions are preserved:

- If F has a *normal*(μ, σ^2) distribution, $\Psi^{\gamma}(F)$ is also a normal distribution with $\mu^* = \mu - \lambda\sigma$ and $\sigma^* = \sigma$,
- If F has a *lognormal*(μ, σ^2) distribution such that $\ln(X) \sim$ *normal*(μ, σ^2), $\Psi^{\gamma}(F)$ is another lognormal distribution with $\mu^* = \mu - \lambda\sigma$ and $\sigma^* = \sigma$.

Let us prove this statement for the case of the lognormal distribution (the case when F has a normal distribution can be easily deducted).

If F has a *lognormal*(μ, σ^2) distribution, then it means that F

$$F(x) = \int_{-\infty}^{x} \frac{1}{\sigma\sqrt{2\pi}} e^{-\frac{(\ln x - \mu)^2}{2\sigma^2}} dx, \tag{A.2}$$

or

$$F(x) = \Phi\left(\frac{\ln x - \mu}{\sigma}\right), \tag{A.3}$$

A. Karimov, *Identifying Stock Market Bubbles*, Contributions to Management Science, https://doi.org/10.1007/978-3-319-65009-8

If we will apply the Wang Transform to the function of F then we will have the following:

$$\begin{aligned}\Psi^{\gamma}(F) &= \Phi(\Phi^{-1}(F)+\gamma) = \Phi\left(\Phi^{-1}\left(\Phi\left(\frac{\ln x-\mu}{\sigma}\right)\right)+\gamma\right) \\ &= \Phi\left(\frac{\ln x-\mu}{\sigma}+\gamma\right) = \Phi\left(\frac{\ln x-(\mu-\gamma\sigma)}{\sigma}\right).\end{aligned} \tag{A.4}$$

The last expression in Eq. (A.4) is the standardization of the lognormal distribution with $\mu^{*} = \mu - \lambda\sigma$ and $\sigma^{*} = \sigma$.

Appendix B
Deriving Bid and Ask Prices for Options Under Brownian Motion Assumptions

B.1 Prices of a European Call Options

The price of the European Call Option C_T can be expressed by the following form:

$$C_T = \max\{(S_T - K), 0\}, \tag{B.1}$$

where S_T is the index price, K is the strike price, and T is the maturity of the option.

B.1.1 Bid Price

By using Eq. (4.4) we can derive the bid price of the European call option:

$$\begin{aligned} b_\gamma(C) &= \int_0^\infty x \, d\Psi^\gamma(F_{C_T}(x)) \\ &= \int_K^\infty (x - K) \, d\Psi^\gamma(F_{S_T}(x)) \\ &= \int_K^\infty x \, d\Psi^\gamma(F_{S_T}(x)) - \int_K^\infty K \, d\Psi^\gamma(F_{S_T}(x)) =: A - B. \end{aligned} \tag{B.2}$$

In order to calculate the part A we will make a variable change of $\ln x = y$, hence the integral limits will change from $\ln K$ to ∞;

$$\begin{aligned} A &= \int_K^\infty x \, d\Psi^\gamma(F_{S_T}(x)) \\ &= \int_{\ln K}^\infty e^y \, d\Phi\left[\frac{y - \ln S_t - \left(r - \frac{1}{2}\sigma^2\right)(T - t) + \gamma\sigma\sqrt{T - t}}{\sigma\sqrt{T - t}}\right] \end{aligned}$$

A. Karimov, *Identifying Stock Market Bubbles*, Contributions to Management Science, https://doi.org/10.1007/978-3-319-65009-8

$$= \int_{\ln K}^{\infty} e^{y} \frac{1}{\sqrt{2\pi}} \exp\left[-\left(\frac{y-\ln S_t-\left(r-\frac{1}{2}\sigma^2\right)(T-t)+\gamma\sigma\sqrt{T-t}}{\sigma\sqrt{T-t}}\right)^2/2\right] d\left(\frac{y}{\sigma\sqrt{T-t}}\right)$$

$$= \int_{\ln K}^{\infty} e^{y} \frac{1}{\sqrt{2\pi}} \frac{1}{\sigma\sqrt{T-t}} \exp\left[-\left(\frac{y-\ln S_t-\left(r-\frac{1}{2}\sigma^2\right)(T-t)+\gamma\sigma\sqrt{T-t}}{\sigma\sqrt{T-t}}\right)^2/2\right] dy$$

$$= \int_{\ln K}^{\infty} \frac{1}{\sqrt{2\pi}} \frac{1}{\sigma\sqrt{T-t}} \exp\left[-\frac{\left(y-\ln S_t-\left(r-\frac{1}{2}\sigma^2\right)(T-t)+\gamma\sigma\sqrt{T-t}\right)^2}{2\sigma^2(T-t)}+y\right] dy. \tag{B.3}$$

Let us take a closer look to the expression in the power of the exponents:

$$\begin{aligned}
A1 &= -\frac{\left(y-\ln S_t-\left(r-\frac{1}{2}\sigma^2\right)(T-t)+\gamma\sigma\sqrt{T-t}\right)^2}{2\sigma^2(T-t)}+y \\
&= -\frac{\left(y-\ln S_t-\left(r-\frac{1}{2}\sigma^2\right)(T-t)+\gamma\sigma\sqrt{T-t}\right)^2-2y\sigma^2(T-t)}{2\sigma^2(T-t)} \\
&= -\frac{\left(y-\left(\ln S_t+\left(r-\frac{1}{2}\sigma^2\right)(T-t)-\gamma\sigma\sqrt{T-t}\right)\right)^2-2y\sigma^2(T-t)}{2\sigma^2(T-t)} \\
&= -\frac{y^2-2y\left(\ln S_t+\left(r-\frac{1}{2}\sigma^2\right)(T-t)-\gamma\sigma\sqrt{T-t}\right)}{2\sigma^2(T-t)} \\
&\quad -\frac{\left(\ln S_t+\left(r-\frac{1}{2}\sigma^2\right)(T-t)-\gamma\sigma\sqrt{T-t}\right)^2-2y\sigma^2(T-t)}{2\sigma^2(T-t)} \\
&= -\frac{y^2-2y\left(\ln S_t+\left(r-\frac{1}{2}\sigma^2\right)(T-t)-\gamma\sigma\sqrt{T-t}+\sigma^2(T-t)\right)}{2\sigma^2(T-t)} \\
&\quad -\frac{\left(\ln S_t+\left(r-\frac{1}{2}\sigma^2\right)(T-t)-\gamma\sigma\sqrt{T-t}\right)^2}{2\sigma^2(T-t)} \\
&= -\frac{y^2-2y\left(\ln S_t+\left(r+\frac{1}{2}\sigma^2\right)(T-t)-\gamma\sigma\sqrt{T-t}\right)}{2\sigma^2(T-t)} \\
&\quad -\frac{\left(\ln S_t+\left(r-\frac{1}{2}\sigma^2\right)(T-t)-\gamma\sigma\sqrt{T-t}\right)^2}{2\sigma^2(T-t)} =
\end{aligned} \tag{B.4}$$

$$= -\frac{y^2 - 2y\left(\ln S_t + \left(r + \frac{1}{2}\sigma^2\right)(T-t) - \gamma\sigma\sqrt{T-t}\right)}{2\sigma^2 (T-t)}$$

$$- \frac{\left(\ln S_t + \left(r - \frac{1}{2}\sigma^2\right)(T-t) - \gamma\sigma\sqrt{T-t} \pm \sigma^2 (T-t)\right)^2}{2\sigma^2 (T-t)}$$

$$= -\frac{y^2 - 2y\left(\ln S_t + \left(r + \frac{1}{2}\sigma^2\right)(T-t) - \gamma\sigma\sqrt{T-t}\right)}{2\sigma^2 (T-t)}$$

$$- \frac{\left(\ln S_t + \left(r + \frac{1}{2}\sigma^2\right)(T-t) - \gamma\sigma\sqrt{T-t} - \sigma^2 (T-t)\right)^2}{2\sigma^2 (T-t)}$$

$$= -\frac{\left(y - \left(\ln S_t + \left(r + \frac{1}{2}\sigma^2\right)(T-t) - \gamma\sigma\sqrt{T-t}\right)\right)^2}{2\sigma^2 (T-t)}$$

$$- \left(\frac{1}{2}\sigma^2 (T-t) - \left(\ln S_t + \left(r + \frac{1}{2}\sigma^2\right)(T-t) - \gamma\sigma\sqrt{T-t}\right)\right)$$

$$= -\frac{\left(y - \left(\ln S_t + \left(r + \frac{1}{2}\sigma^2\right)(T-t) - \gamma\sigma\sqrt{T-t}\right)\right)^2}{2\sigma^2 (T-t)} + \ln S_t + r(T-t) - \gamma\sigma\sqrt{T-t}. \tag{B.5}$$

Putting the last expression back to the integral gives us:

$$A = \int_{\ln K}^{\infty} \frac{1}{\sqrt{2\pi}} \frac{1}{\sigma\sqrt{T-t}} \exp\left[-\frac{\left(y - \left(\ln S_t + (r + \frac{1}{2}\sigma^2)(T-t) - \gamma\sigma\sqrt{T-t}\right)\right)^2}{2\sigma^2 (T-t)}\right]$$

$$\times \exp\left(\ln S_t + r(T-t) - \gamma\sigma\sqrt{T-t}\right) dy$$

$$= \int_{\ln K}^{\infty} \frac{1}{\sqrt{2\pi}} \exp\left(-\left(\frac{y - \left(\ln S_t + (r + \frac{1}{2}\sigma^2)(T-t) - \gamma\sigma\sqrt{T-t}\right)}{\sigma\sqrt{T-t}}\right)^2 / 2\right)$$

$$\times \exp\left(\ln S_t + r(T-t) - \gamma\sigma\sqrt{T-t}\right) d\frac{y}{\sigma\sqrt{T-t}}$$

$$= \exp\left(\ln S_t + r(T-t) - \gamma\sigma\sqrt{T-t}\right)$$

$$\times \int_{\ln K}^{\infty} \frac{1}{\sqrt{2\pi}} \exp\left(-\left(\frac{y - \left(\ln S_t + (r + \frac{1}{2}\sigma^2)(T-t) - \gamma\sigma\sqrt{T-t}\right)}{\sigma\sqrt{T-t}}\right)^2 / 2\right) d\frac{y}{\sigma\sqrt{T-t}}$$

$$= e^{\ln S_t + r(T-t) - \gamma\sigma\sqrt{T-t}} \cdot \Phi\left[\frac{\ln S_t + (r + \frac{1}{2}\sigma^2)(T-t) - \gamma\sigma\sqrt{T-t} - \ln K}{\sigma\sqrt{T-t}}\right]. \tag{B.6}$$

Basically the first integral in Eq. (B.2) will have the following expression:

$$A = \int_K^{\infty} x \, d\Psi^{\gamma}(F_{S_T}(x))$$
$$= e^{\ln S_t + r(T-t) - \gamma\sigma\sqrt{T-t}} \cdot \Phi\left[\frac{\ln \frac{S_t}{K} + (r + \frac{1}{2}\sigma^2)(T-t) - \gamma\sigma\sqrt{T-t}}{\sigma\sqrt{T-t}}\right],$$

and the second integral can be calculated as:

$$B = \int_K^{\infty} K \, d\Psi^{\gamma}(F_{S_T}(x)) = K\left(1 - \Phi\left[\frac{\ln \frac{K}{S_t} - \left(r - \frac{1}{2}\sigma^2\right)(T-t) + \gamma\sigma\sqrt{T-t}}{\sigma\sqrt{T-t}}\right]\right)$$
$$= K \cdot \Phi\left[\frac{\ln \frac{S_t}{K} + \left(r - \frac{1}{2}\sigma^2\right)(T-t) - \gamma\sigma\sqrt{T-t}}{\sigma\sqrt{T-t}}\right].$$

The following transformation have been used in the derivation of the part B:

$$1 - F_{S_T}(x) = 1 - \Phi\left[\frac{\ln \frac{x}{S_t} - \left(r - \frac{1}{2}\sigma^2\right)(T-t)}{\sigma\sqrt{T-t}}\right] = \Phi\left[\frac{\ln \frac{S_t}{x} + \left(r - \frac{1}{2}\sigma^2\right)(T-t)}{\sigma\sqrt{T-t}}\right].$$

One of the assumptions of the Black-Scholes Model is that the starting premium is borrowed at the risk-free rate, therefore, we will need to use a suitable discount factor. Hence, the price needs to be multiplied by discount factor $e^{-r(T-t)}$, which leads to the following expression for the bid price:

$$b_{\gamma}(C) = S_t e^{-\gamma\sigma\sqrt{T-t}} \cdot \Phi\left[\frac{\ln \frac{S_t}{K} + (r + \frac{1}{2}\sigma^2)(T-t) - \gamma\sigma\sqrt{T-t}}{\sigma\sqrt{T-t}}\right]$$
$$- e^{-r(T-t)} K \cdot \Phi\left[\frac{\ln \frac{S_t}{K} + \left(r - \frac{1}{2}\sigma^2\right)(T-t) - \gamma\sigma\sqrt{T-t}}{\sigma\sqrt{T-t}}\right].$$

B.1.2 Ask Price

By using Eq. (4.3) we can derive the ask price of the European call option:

$$a_{\gamma}(C) = -\int_{-\infty}^{0} x \, d\Psi^{\gamma}(F_{-C_T}(x))$$
$$= -\int_{-\infty}^{0} x \, d\Psi^{\gamma}(1 - F_{S_T}(K - x))$$

$$
\begin{aligned}
&= -\int_0^{\infty} x \, d\Psi^{\gamma}(1 - F_{S_T}(K + x)) \\
&= -\int_K^{\infty} (x - K) \, d\Psi^{\gamma}(1 - F_{S_T}(x)) \\
&= -\int_K^{\infty} x \, d\Psi^{\gamma}(1 - F_{S_T}(x)) + \int_K^{\infty} K \, d\Psi^{\gamma}(1 - F_{S_T}(x)) =: -A + B.
\end{aligned} \tag{B.7}
$$

By employing the same steps which we used during the calculation of bid price we can derive the first integral in Eq. (4.17) as follows:

$$
\begin{aligned}
A &= \int_K^{\infty} x \, d\Psi^{\gamma}(1 - F_{S_T}(x)) \\
&= e^{\ln S_t + r(T-t) + \gamma\sigma\sqrt{T-t}} \cdot \Phi\left[\frac{\ln \frac{S_t}{K} - (r + \frac{1}{2}\sigma^2)(T-t) + \gamma\sigma\sqrt{T-t}}{\sigma\sqrt{T-t}}\right],
\end{aligned}
$$

and the second integral can be calculated as:

$$
\begin{aligned}
B = \int_K^{\infty} K \, d\Psi^{\gamma}(1 - F_{S_T}(x)) &= K \cdot \left(1 - \Phi\left[\frac{\ln \frac{K}{S_t} + \left(r - \frac{1}{2}\sigma^2\right)(T-t) - \gamma\sigma\sqrt{T-t}}{\sigma\sqrt{T-t}}\right]\right) \\
&= K \cdot \Phi\left[\frac{\ln \frac{S_t}{K} - \left(r - \frac{1}{2}\sigma^2\right)(T-t) + \gamma\sigma\sqrt{T-t}}{\sigma\sqrt{T-t}}\right].
\end{aligned}
$$

By combining the part of A and B and multiplying the price with the discount factor of $e^{-r(T-t)}$ will leads to the following expression for the bid price:

$$
\begin{aligned}
a_{\gamma}(C) = S_t e^{\gamma\sigma\sqrt{T-t}} \cdot \Phi\left[\frac{\ln \frac{S_t}{K} + (r + \frac{1}{2}\sigma^2)(T-t) - \gamma\sigma\sqrt{T-t}}{\sigma\sqrt{T-t}}\right] \\
- e^{-r(T-t)} K \cdot \Phi\left[\frac{\ln \frac{S_t}{K} - \left(r - \frac{1}{2}\sigma^2\right)(T-t) + \gamma\sigma\sqrt{T-t}}{\sigma\sqrt{T-t}}\right].
\end{aligned}
$$

B.2 Prices of a European Put Options

The price of the European Put Option can be expressed by the following function:

$$
P_T = \max\{(K - S_T), 0\}, \tag{B.8}
$$

where S is the index price, K is the strike price, and T is the maturity of the option.

B.2.1 Bid Price

By using Eq. (4.4) and also the transformations that we used in the process of calculation of bid and ask prices of the European call option we can derive the bid price of the European put option as follows:

$$\begin{aligned}
b_\gamma(P) &= \int_0^\infty x\, d\Psi^\gamma(F_{P_T}(x)) \\
&= \int_0^\infty x\, d\Psi^\gamma(1 - F_{S_T}(K - x)) = -\int_0^K (K - x)\, d\Psi^\gamma(1 - F_{S_T}(x)) \\
&= -\int_0^K (K - x)\, d\Psi^\gamma \left(1 - \Phi\left(\frac{\ln \frac{x}{S_t} - \left(r - \frac{1}{2}\sigma^2\right)(T - t)}{\sigma\sqrt{T - t}}\right)\right) \\
&= -\int_0^K (K - x)\, d\Psi^\gamma \left(\Phi\left(\frac{\ln \frac{S_t}{x} + \left(r - \frac{1}{2}\sigma^2\right)(T - t)}{\sigma\sqrt{T - t}}\right)\right) \\
&= -\int_0^K (K - x)\, d\Phi\left(\frac{\ln \frac{S_t}{x} + \left(r - \frac{1}{2}\sigma^2\right)(T - t) + \gamma\sigma\sqrt{T - t}}{\sigma\sqrt{T - t}}\right) \\
&= \int_0^K (K - x)\, d\Phi\left(\frac{\ln \frac{x}{S_t} - \left(r - \frac{1}{2}\sigma^2\right)(T - t) - \gamma\sigma\sqrt{T - t}}{\sigma\sqrt{T - t}}\right).
\end{aligned} \tag{B.9}$$

Again we split the integral into two parts and afterwards apply the discount factor in order to get the formula of the bid price of European put option:

$$\begin{aligned}
b_\gamma(P) = e^{-r(T-t)} K \Phi &\left[\frac{\ln \frac{K}{S_t} - \left(r - \frac{1}{2}\sigma^2\right)(T - t) - \gamma\sigma\sqrt{T - t}}{\sigma\sqrt{T - t}}\right] \\
- S_t e^{\gamma\sigma\sqrt{T-t}} \cdot \Phi &\left[\frac{\ln \frac{K}{S_t} - (r + \frac{1}{2}\sigma^2)(T - t) - \gamma\sigma\sqrt{T - t}}{\sigma\sqrt{T - t}}\right].
\end{aligned}$$

B.2.2 Ask Price

By using Eq. (4.3) and also the transformations that we used in the process of calculation of bid and ask prices of the European call option we will derive the ask price of the European put option as follows:

$$\begin{aligned}
a_\gamma(P) &= -\int_{-\infty}^0 x\, d\Psi^\gamma(F_{-P_T}(x)) \\
&= -\int_{-\infty}^0 x\, d\Psi^\gamma(F_{S_T}(K + x))
\end{aligned}$$

$$= -\int_0^{\infty} x \, d\Psi^{\gamma}(F_{S_T}(K-x)) \tag{B.10}$$

$$= \int_0^{K} (K-x) \, d\Psi^{\gamma}(F_{S_T}(x))$$

$$= \int_0^{K} (K-x) \, d\Phi\left(\frac{\ln\frac{x}{S_t} - \left(r - \frac{1}{2}\sigma^2\right)(T-t) + \gamma\sigma\sqrt{T-t}}{\sigma\sqrt{T-t}}\right).$$

Again we split the integral into two parts and afterwards apply the discount factor in order to get the formula of the ask price of European put option

$$a_\gamma(C) = e^{-r(T-t)} K\Phi\left[\frac{\ln\frac{K}{S_t} - \left(r - \frac{1}{2}\sigma^2\right)(T-t) + \gamma\sigma\sqrt{T-t}}{\sigma\sqrt{T-t}}\right]$$

$$- S_t e^{-\gamma\sigma\sqrt{T-t}} \cdot \Phi\left[\frac{\ln\frac{K}{S_t} - (r + \frac{1}{2}\sigma^2)(T-t) + \gamma\sigma\sqrt{T-t}}{\sigma\sqrt{T-t}}\right].$$

In Table B.1 we provide closed form formulas for the bid and ask prices of the European call and put options:

Table B.1 Bid and ask prices of European options

Option	Price	d_1	d_2
b_γ (C)	$S_t e^{-\gamma\sigma\sqrt{T-t}}\Phi(d_1) - e^{-r(T-t)}K\Phi(d_2)$	$\frac{\ln\frac{S_t}{K}+(r+\frac{1}{2}\sigma^2)(T-t)-\gamma\sigma\sqrt{T-t}}{\sigma\sqrt{T-t}}$	$d_1 - \sigma\sqrt{T-t}$
a_γ (C)	$S_t e^{\gamma\sigma\sqrt{T-t}}\Phi(d_1) - e^{-r(T-t)}K\Phi(d_2)$	$\frac{\ln\frac{S_t}{K}-(r+\frac{1}{2}\sigma^2)(T-t)+\gamma\sigma\sqrt{T-t}}{\sigma\sqrt{T-t}}$	$d_1 - \sigma\sqrt{T-t}$
b_γ (P)	$e^{-r(T-t)}K\Phi(d_2) - S_t e^{\gamma\sigma\sqrt{T-t}}\Phi(d_1)$	$\frac{\ln\frac{K}{S_t}-(r+\frac{1}{2}\sigma^2)(T-t)-\gamma\sigma\sqrt{T-t}}{\sigma\sqrt{T-t}}$	$d_1 + \sigma\sqrt{T-t}$
a_γ (P)	$e^{-r(T-t)}K\Phi(d_2) - S_t e^{-\gamma\sigma\sqrt{T-t}}\Phi(d_1)$	$\frac{\ln\frac{K}{S_t}-(r+\frac{1}{2}\sigma^2)(T-t)+\gamma\sigma\sqrt{T-t}}{\sigma\sqrt{T-t}}$	$d_1 + \sigma\sqrt{T-t}$

MIX
Papier aus verantwortungsvollen Quellen
Paper from responsible sources
FSC® C105338

If you have any concerns about our products,
you can contact us on
ProductSafety@springernature.com

In case Publisher is established outside the EU,
the EU authorized representative is:
Springer Nature Customer Service Center GmbH
Europaplatz 3, 69115 Heidelberg, Germany

Printed by Libri Plureos GmbH
in Hamburg, Germany